Femtosecond Laser–Matter Interactions

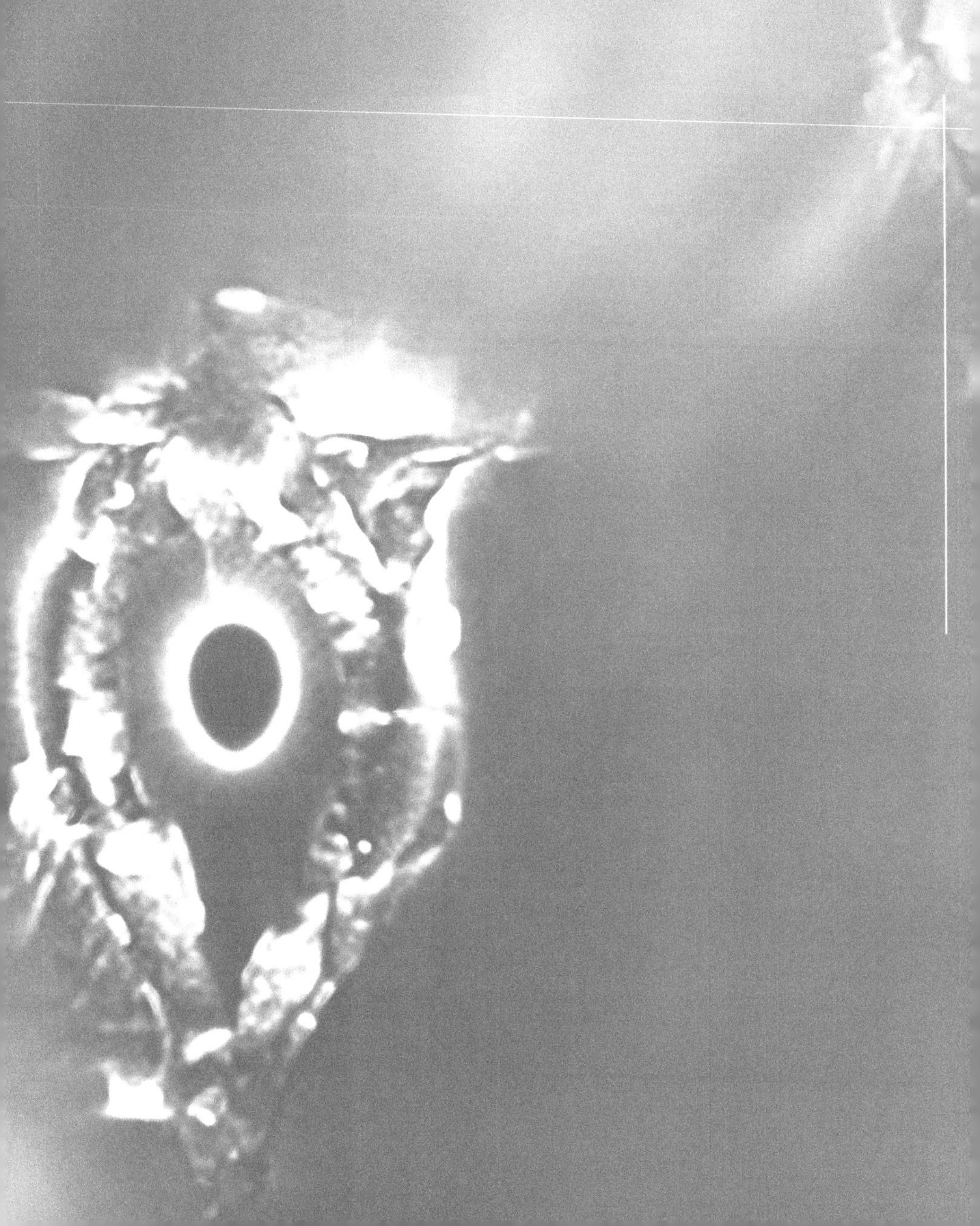

Femtosecond Laser–Matter Interactions

Solid-Plasma-Solid Transformations at the Extreme Energy Density

Eugene G. Gamaly

CRC Press
Taylor & Francis Group
Boca Raton London New York

CRC Press is an imprint of the
Taylor & Francis Group, an **informa** business

Femtosecond Laser-Matter Interactions: Solid-Plasma-Solid Transformations at the Extreme Energy Density

First published 2022 by Jenny Stanford Publishing Pte. Ltd.

Published 2023 by CRC Press
Taylor & Francis Group
6000 Broken Sound Parkway NW, Suite 300
Boca Raton, FL 33487-2742

ISBN 13: 978-981-4877-40-4 (hbk)

Visit the Taylor & Francis Web site at
http://www.taylorandfrancis.com

and the CRC Press Web site at
http://www.crcpress.com

British Library Cataloguing-in-Publication Data
A catalogue record for this book is available from the British Library.

To Ksana

Contents

Preface

The aim of this book is to present a succinct description of *the non-equilibrium physics* of the laser–solid interactions induced by the action of the ultra-short laser pulse on metals and dielectrics. The ultra-short pulse stands for the pulses shorter of the major relaxation times in the laser-excited solids (in the sub-picosecond time domain). Such a pulse excites only electrons while the atoms remain at the initial temperature and at the initial positions. We consider interactions in a broad range of intensities from the subtle excitation of the coherent phonons at low intensity up to the creation of the extreme energy density by the intense short laser focused into sub-micron volume.

The basic physics of the interaction is the focus of the studies. Therefore, we present mainly the analytical estimates in a form allowing to a reader using (or creating) the numerical code on a clear physical basis. The effects of the laser field on the matter and the effects of the excited matter on the field are considered intertwined.

The goal of the book is the study of the interaction of the intense ultra-short laser pulse focused to the sub-micron volume inside the transparent dielectric at the laboratory tabletop. The interaction zone and laser-affected material remain confined inside the pristine crystal. The absorbed energy density exceeds a few TPa (1 TPa = 10^6 J/cm^3 = 10 Mbar) that is higher than Young's modulus of any existing material. Hence, the micro-explosion after the end of the pulse creates a void inside an undamaged crystal, the size of which can be controlled by the proper choice of the laser and material parameters. It was found that in the conditions of micro-explosion the unusual high-pressure phase transformations occurred resulting in the formations of bcc aluminium and novel phases of silicon.

The phase transitions in the laser-induced confined micro-explosion take place along the unique transformation path first transforming solid into the chaotic state where the memory of initial structure is lost. Transition from the randomised state to the ambient conditions occurs allegedly along the path where the kinetic barriers are suppressed resulting in the previously unobserved structures.

The laser intensity during the interaction increases up to few hundreds of TW/cm^2 well below the relativistic intensity. We describe the changes in the material properties step by step as the intensity grows up to the maximum. We provide the reader a succinct description of the ultra-short laser–matter interaction starting from the low intensities, when a solid remains intact after coming back to the ambient state after the interaction. Step by step, we proceed up to the high intensities when any solid is converted to plasma and the extreme energy density is achieved. The analytical formulae, scaling relations, describing the major characteristics of laser–matter interaction are combinations of the laser and material parameters having a clear physical meaning. These relations allow planning the experiments; interpret the results, indicating the methods for control over the phenomena. The rigorous formulation of the problems for the computer simulations naturally follows from this approach.

Energy transfer from electrons to atoms (ions) occurs after the end of the pulse producing on the route to equilibrium the wide variety of unusual transient material states. At low intensity, the laser light produces the subtle material modification keeping material intact. The low-intensity laser-induced transformations are reversible: A material returns to the initial state after cooling to the ambient conditions. The laser can excite short-lived coherent atomic vibrations (phonons). At elevated intensity, laser excitation creates the transient states with the properties, which cannot be attributed to either solid or liquid, metal or dielectric.

The tightly focused short laser pulse may achieve the maximum intensity of hundreds TW/cm^2 at the focal spot generating the energy density of a few MJ/cm^3 (pressure a few TPa) in the focal volume confined inside of a transparent crystal. Any material in this volume is converted to the hot solid density plasma. The memory of the initial structure is lost in this high entropy state. The return to the ambient condition occurs along the unique chaos-to-equilibrium routes drastically different from the conventional high-pressure transitions. The novel, never observed previously, high-pressure phases of aluminium (bcc-Al) and silicon were discovered under these conditions.

We limit the consideration of ultra-short laser/matter interaction mainly to a single-pulse interaction with solids, metals and dielectrics. The chain of the inter-connected non-equilibrium processes in a single pulse-solid interaction and assumptions made are explicitly presented. This is the first step to the description of the material modification by the multiple-pulse action on the same spot. This problem is much more complicated and less studied experimentally and theoretically and not considered in this book. However, the single-pulse ablation studies were successfully extended to the material ablation under the multiple-pulse action. The theoretical and experimental studies of the material ablation by the high-repetition-rate lasers (up to 96 MHz), described in the book, demonstrated that it is possible to control and predict the multiple-pulse ablation process.

The short/intense laser–matter interactions found numerous applications in industry (micro-machining), information technology (optical memory, waveguides, photonic crystals), materials science (new material, nano-cluster and nano-structure formation), biology (selective action on cells and big molecules) and medicine (dentistry and surgery). In this book, we concentrate on achieving the extreme energy density and creation of conditions where the unique paths for the formation of the novel high-pressure material phases are created.

The knowledge of physics in the frames of the university course of physics is sufficient for the understanding and practical use of this book. We believe the book is accessible for advanced undergraduate and graduate students, practitioners and newcomers to the field, either theorists or experimentalists.

This book is the updated version of the book Gamaly (2011) partly on the basis of the review articles (Gamaly, 2011; Gamaly and Rode, 2013) and on the lectures (Aix-Marseille University, 2014, 2015; the Photonic Centre of Osaka University, 2019). The book is neither a review of the author's works nor a review of the works of numerous groups working in applications of the ultra-short laser–matter interactions in micro-machining, writing, cutting and welding. The book is dedicated to a sufficiently complete description of the *non-equilibrium* physics of the ultra-short laser–solid interactions, which is the basis for any application.

I am indebted to my colleagues who helped to understand many phenomena in the course of the joint work on the projects described in the book. I would like to express my sincere gratitude to Andrei Rode, Barry Luther-Davies, Wieslaw Krolikowski (Australian National University), Saulius Juodkazis (Swinburne University), Olivier Uteza, and Marc Sentis (Aix-Marseille University) for the many enlightening discussions of the books topics. My special gratitude to Professor Vladimir Tikhonchuk (University of Bordeaux), who read the manuscript and made numerous comments, which were implemented in the revised text. My sincere gratitude to Professor Juodkazis for the kind permission for using his photograph of the void in sapphire for the front cover.

The list of the references reflects the author's views on the most important papers and does not cover the entire field completely. The author considers the cited papers as the guides in the vast sea of publications.

The ultra-short laser–matter interactions studies are multidisciplinary by nature. The topics include optics, the electromagnetic field interaction with solids and plasma, atomic physics, fundamentals of plasma physics, elements of the solid-state physics, and statistical physics with the glimpses from the quantum mechanics. In order to account these topics succinctly, the book has been organised as follows.

The first chapter is dedicated to the description of the main features of the light source as the interaction tool and the characteristics of a matter response to the laser action. The second chapter considers the laser interaction with metals. The third chapter is devoted to the laser interaction with dielectrics up to the conversion of a dielectric to plasma. The unconventional transient material states created at the gradually increasing laser intensity are described in the fourth chapter. The fifth chapter considers the processes of the non-equilibrium evaporation (ablation) of metals and dielectrics under the action of the intense ultra-short laser beam. Sixth chapter describes the ways for the light concentration inside a transparent dielectrics allowing the achievement of the extreme energy density and preserve the pressure/temperature-affected material returned to the ambient being confined inside a pristine crystal. The methods and processes leading to the extreme energy density are also considered in Chapter 6 along with the experimental results that evidence the formation of previously unobserved novel material phases. In the conclusive remarks, we address the unresolved problems and future directions in the ultra-short laser–matter interactions studies.

We demonstrate that the understanding of the fundamental features of the laser–matter interactions phenomena is the direct way to the technical solutions allowing the implementation of academic studies' results into the applications such as micro-machining and new material formation. We present the description of the relevant experiments, experimental setups, experimental results and comparison to the theoretical predictions.

In the book, we are using the Gaussian (CGSE) system of units since the SI units are incongruous for the description of the electromagnetic phenomena. The dimensions of the electric and magnetic field strengths are different in SI, even though both enter the Maxwell equations (and the electromagnetic field tensor) in a quite equivalent manner. The table for the conversion of CGSE units into SI units is presented at the end of the book.

List of Frequently Used Symbols

γ:	ratio of specific heats (adiabatic constant)
ΔH_f:	enthalpy of fusion (melting)
ΔH_{vac}:	enthalpy of point defect formation
ΔS_f:	entropy of fusion (melting)
ε:	dielectric permittivity
ε_b:	binding (cohesive) energy of atoms
ε_c:	"cold" specific energy
ε_F:	Fermi energy
ε_{osc}:	energy of free electron oscillations in the high frequency electric field
λ:	laser wavelength
ν_{ei}:	electron-ion momentum/energy exchange rates
$\nu_{\text{e-ph}}$:	electron-phonon momentum/energy exchange rates
σ:	conductivity
τ_{las}, t_p:	laser pulse duration
ω, ω_0:	laser frequency
$\omega_{\text{pe}}, \omega_{\text{pi}}$:	electrons and ions plasma frequency
ω_{ph}:	phonon's frequency
$A = E_{\text{abs}}/E_{\text{las}}$:	absorption coefficient
$C_{i\text{-}}, C_e, C_L$:	specific heat of ion, electrons and lattice
D:	diffusion coefficient
d:	inter-atomic spacing (closest neighbours distance)
E:	amplitude of the electric field
E_{abs}:	absorbed energy density
$F = E_{\text{las}}/S_{\text{foc}}$:	laser energy density per unit area at the focal spot (fluence)
I, I_0:	laser intensity (energy per unit surface per unit time)
J:	ionization potential
l_{De}:	Debye screening length
l_s:	skin-layer thickness (normal skin-effect)
M:	atom (ion) mass

m_e: free electron mass
n,κ: real and imaginary parts of the refractive index
n_a: atomic number density
n_c: critical number density of electrons
N_{D}: number of particles in the Debye sphere
n_e: number density of electrons
p_{cold}: "cold" elastic pressure
q: amplitude of phonons
R: Fresnel reflection coefficient
S_{foc}: focal spot area
T_e: electron temperature
T_i: ion temperature
T_L: lattice temperature
T_m: melting temperature
v_{sound}: speed of sound
w_{esc}: electronic work function for metals
Y: Young modulus

Chapter 1

Basics of Laser–Matter Interactions: Light and Matter

In this chapter, we identify different ultra-short laser–matter interaction regimes. These regimes are separated by the intensity, pulse duration, nature of a target (transparent or opaque solid) beam position on the target (on the surface, inside a bulk, at rear surface) and the focussing conditions. As a first step, we present the description of the major tool, laser beam, and the studied subject—metal and dielectric solids.

1.1 Laser Beam

It is instructive to describe the spatial, temporal and internal structure of the incident laser beam. Let us characterize first the unfocussed beam and approximations used for the characterization of the beam in practice.

1.1.1 Macroscopic Electrodynamics

From the quantum viewpoint, a laser beam is the flow of photons. The photons are massless relativistic particles with the energy

Femtosecond Laser-Matter Interactions: Solid-Plasma-Solid Transformations at the Extreme Energy Density
Eugene Gamaly

ISBN 978-981-4877-40-4 (Hardcover), 978-1-003-25661-8 (eBook)
www.jennystanford.com

$\hbar\omega$ ($\hbar$ is the Planck constant divided by two π, and ω is the cyclic frequency). The energy flow of photons per unit surface per unit time, intensity, expressed through the photons number density, n_{ph} [cm^{-3}], moving with the speed of light in vacuum, c, reads

$$I = n_{\text{ph}} \cdot \hbar\omega \cdot c. \tag{1.1}$$

The photon's wavelength relates to the frequency by the familiar relation, $\lambda = 2\pi c/\omega$. Quantum mechanical description of photons beam is necessary when the number density of photons is less than one photon per cubic wavelength (Landau et al., 1984). One can describe the energy flow expressed through the electric field amplitude in the frames of macroscopic electrodynamics, if the number density of photons is large, $n_{\text{ph}} >> \lambda^{-3}$ ($\sim 10^{12}$ photons/cm^3 for $\lambda = 1$ μm). Expressing the number density through the intensity, $n_{\text{ph}} = I/\hbar \times c \times \omega$, one can find that even for low intensity of interest for the material modification, $\sim 10^9$ W/cm^2, the density of photons exceeds 2×10^{17} photons/cm^3, well over the above limit. Another condition of applicability of macroscopic electrodynamics demands that the field penetration depth into a medium, l_s, should be significantly larger than the inter-atomic distance, $l_s >> (n_a)^{-1/3}$, where n_a is the atomic number density. This condition is fulfilled in the all considered situations. Thus, we can use the classical description of the incident photons' beam in the frame *of macroscopic electrodynamics* for considered below interaction regimes.

1.1.2 Polarization States

Photon as a quantum particle possesses spin equal to one (in units of the Plank constant), $s = 1$. In accord with the quantum mechanics, the particle possessing spin, s, can be found in $(2s + 1)$ states. In classical electrodynamics, the three spin states of photon correspond to three polarization states. Three quantum states of photon, corresponding to the spin states ±1, 0, are retained in classical description as polarization states corresponding to the left hand and right hand circular polarization ($\alpha = \pm 1$) and linear polarization ($\alpha = 0$).

Slow varying envelope. We will use in this book mainly the slow varying envelope approximation when the major time scale (pulse duration) significantly exceeds laser period ($t >> T/2\pi$). The electric field then is expressed as follows: $E = E_0(t)(e_x \cdot \cos\omega t \pm a \cdot e_y \cdot \sin\omega t)$. Here e_x, e_y are the unit vectors in a plane perpendicular to the wave vector of the wave propagating in the z-direction.

The squared laser field averaged over the many laser periods reads

$$\langle E^2 \rangle = E^2(t) \cdot \int^{t>>2\pi/\omega} [\cos\omega\tau + \alpha \sin\omega\tau]^2 d\tau \cdot \quad (1.2)$$

The laser intensity, I (W/cm^2), takes the form (brackets indicating averaging are omitted for brevity)

$$I = \frac{cE^2}{8\pi}(\alpha^2 + 1). \quad (1.3)$$

1.1.3 Spectral Structure

The initial spectral structure of the pulse assumed to be well defined with the central frequency and the band's width corresponding to the pulse duration. For the transform-limited pulse (Akhmanov et al., 1988), the bandwidth and the pulse duration shall obey the condition, $\Delta\omega \times t_p \leq 1$. For example for the pulse with t_p = 100 fs, λ = 800 nm, one obtains, $\Delta\lambda \approx (\lambda\Delta\omega/\omega) \approx 3.39$ nm. There are no theoretically predicted and experimentally confirmed interaction effects of the initial beam spectral structure on the interaction with solids to the best of our knowledge. However, in the interaction of intense ionizing laser beam with gases (at intensity $\geq 10^{14}$ W/cm^2), it was established that under ionization the central frequency experienced the blue shift and pulse duration was shortened (Wood et al., 1988).

If the pulse duration is less than recombination time and time for the escape of electrons from the focus then the dielectric permeability of laser-affected gas depends on the fast rising electron density. The ionization rate varies on the time scale comparable with the laser period. Therefore, non-linear current varies on the same time scale affecting time, space and spectral characteristics of the propagating pulse and resulting in blue shift of central frequency (on 10–18 nm) and change in the group velocity of the pulse.

1.1.4 Temporal/Spatial Shape of the Incident Laser Pulse Intensity

The pulse intensity dependence on time and on space across the focal spot can be presented in a general form:

$$I(r,t) = f(r) \cdot i(t)\,[\mathrm{W/cm^2}] \tag{1.4}$$

The sophisticated spatial and temporal variations of the laser pulse (structured light) are beyond the scope of this book. We start with the conventional spatial distribution of intensity across the circular beam of radius, r_f, which has the shape close to the Gaussian:

$$f(r) = I_m \cdot \exp(-2r^2/r_f^2) \tag{1.5}$$

The time dependence during the pulse duration of t_p often is taken in the form of the Gauss, $i(t) = \exp(-2t^2/t_p^2)$. The energy absorbed per unit area (fluence) in any spatial point reads

$$F(r,t) = I_m \cdot \exp(-2r^2/r_f^2) \int_0^t i(\tau)\,d\tau. \tag{1.6}$$

Thus, in any spatial spot across the focal area the effect of the laser beam directly depends on the absorbed laser fluence. In order to keep the laser–matter interaction strictly within the short pulse duration time the laser energy should be delivered during this period. Therefore, the important condition of keeping the interaction in the ultra-short mode is elimination of any pre- or after pulses maintaining high-energy contrast. The contrast is the ratio of the maximum fluence in the main pulse to the fluence in the pre-pulse. Fluence at the pre-pulse should be kept much lower than the ionization threshold for dielectrics (see Chapter 3), thus keeping a target close to the initial state and not allowing pre-plasma formation, which may completely change the interaction mode.

The criterion of the short pulse/matter interaction mode is the contrast ratio of the intensity in the main pulse to the intensity of the pre-pulse. Thorough analysis of the necessary contrast ratio has been provided by Luther-Davies et al. (1992). We assume that the contrast ratio should be $I_{\text{main}}/I_{\text{pre-pulse}} > 10^{+9}$.

The ideal pulse shape in space for the realization of the ultra-short interaction regime is the pulse of square form ("flat-top-hat" pulse) (see Fig. 1.1). In this case, the whole laser affected material across the focal spot would be in the same state. We will discuss the interactions of the laser beams with the different spatial intensity distributions across the beam (the Bessel beams) later in the book.

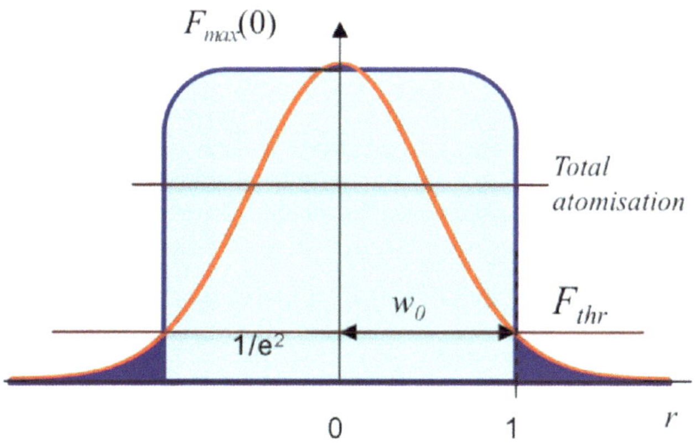

Figure 1.1 The spatial shape of the laser pulse: pink line: the Gauss pulse; blue line: flat-top-hat shaped pulse.

1.1.5 Focussing Positions of the Laser Beam on/in a Target

Laser–matter interactions are performed in the different relative positions of the tool (laser) and material (target). Experiments were performed when a laser focusses at the opaque surface of the target (either in air or in vacuum), inside the bulk of a transparent target (crystal or liquid) at the back surface of the transparent target (light-induced forward transfer (LIFT) mode, see Zegriotti et al., 1998) or at the surface of the opaque layer buried under the transparent layer (see Fig. 1.2).

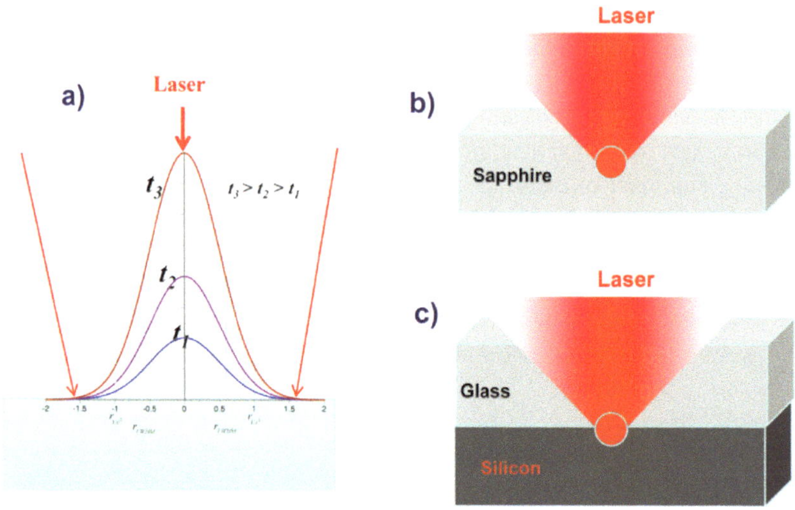

Figure 1.2 Focussing laser beam (a) on the surface of a target (red lines are the spatial intensity distributions across the focal spot at the different time moments); (b) in the bulk of a transparent solid; (c) at the surface of opaque solid buried under the transparent layer.

1.1.6 Focussing to Diffraction Limit with the High NA Lens

The minimum space scale of the volume where the photons may be concentrated follows from the uncertainty relation in quantum mechanics, $\Delta p \cdot \Delta x \geq \hbar/2$. Taking for the photon's momentum, $\Delta p \approx \hbar\omega/c$, one immediately gets a minimum scale to be a fraction of the wavelength, $\Delta x \geq c/2\omega \approx \lambda/4\pi$.

One can show how the focusing to diffraction limit can be achieved in practice. A plane wave focused by a lens creates the axially symmetric intensity distribution at the focus (Born and Wolf, 2003). The intensity distribution in the focal plane is taken the same as for diffraction of a plane wave on a round aperture. We take the radius of the first Airy disk, r_f, as a focal spot radius. The focal radius depends on the focusing angle, θ (a half of the focusing cone angle, see Fig. 1.3), and on the refractive index at the focus, n. The numerical aperture of the lens is customarily defined as, $\mathrm{NA} = n \times \sin\theta$. Then the focal spot radius is given by

$$r_f = 0.61\,\lambda/\mathrm{NA} \tag{1.7}$$

Note that due to Snell's law, $n_{\text{air}}\sin\theta_{\text{air}} = n_m \sin\theta_m$, the NA is the same for focusing inside material or in air $n_{\text{air}} = 1$. The focused Gaussian beam has no minimum in the focal intensity distribution. The energy enclosed within the Airy disk constitutes 83% of the incident energy. 86% of total energy of the Gauss pulse is contained within a radius taken at intensity equal to $1/e^2$ of the maximum value. Thus, it is reasonable to take the focal spot size diameter as that for the first Airy disk. The length of the focal region is conventionally taken as the doubled Rayleigh length:

$$z_R = \pi r_f^2 n/\lambda \tag{1.8}$$

It should be noted that this expression is approximate, since the actual laser beams are neither plane waves nor Gaussian.

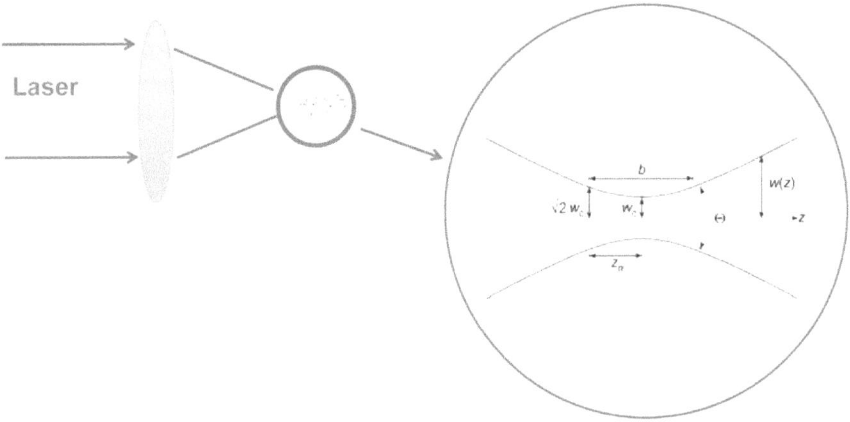

Figure 1.3 Scheme of the intensity distribution in the tight focusing in the bulk of a transparent solid.

*Modification in high-*NA *focusing.* It was demonstrated in several papers that in the focus of high-NA objective, where paraxial approximation fails, the internal structure of the beam drastically changes. Neminen et al. (2008) claimed that strong focussing even converts the circularly polarized Gaussian beam into the longitudinal vortex possessing the topological phase, phase singularity and axial component of the electric field.

Bomzon et al. (2006) showed that the energy distribution in the focal plane of a beam focused with high-NA lens possesses a shape, which depends on polarization. A focal spot from circularly polarized light has a circular symmetry in contrast to the asymmetric distribution observed when linearly polarized light is focused with the lens of NA = 0.95. Focusing a beam through a high-NA objective also causes space-varying polarization and depolarization. They found the threshold at which the increase in NA starts changing the amount of the geometrical phase in the beam. The ratio between the angular momentum and the energy of the entire beam might be considered as a portion of the beam possessing the Berry phase. This ratio is almost constant when the NA is smaller than 0.2, after which it monotonically increases to a value of about 1.4 at NA = 0.95. The authors explained this apparent paradox showing that the lens acts as a low pass filter. The lens only transmits spatial frequencies lower than a cut-off wave number. The high spatial frequencies carry more angular momentum, related to the geometrical phase, per unit energy than the low spatial frequencies. The cut-off wave number increases with NA. Therefore, increasing the aperture and hence the NA inevitably increases the amount of optical angular momentum (OAM) per unit energy in the focal region. It should be noted that these phenomena are not understood fully and need more theoretical and experimental studies.

1.1.7 Diffraction-Free Beams

Diffraction (uncertainty principle) sets the low limit for the focal spot size and therefore on the maximum energy density achievable at the tight focusing. Moreover, the focal volume has also a limit set by the Rayleigh range that is also proportional to the focal spot radius.

Three decades ago, Durnin was found the solution of the Helmgoltz equation (see Durnin, 1987-1) describing the monochromatic light beam having specific intensity distribution across the beam, $J_0^2(k_r r)$, which propagates in the z-direction keeping the across-the-beam intensity distribution unchanged. Here J_0 is the zeroth-order Bessel function of the first kind.

The ideal diffraction-free Bessel beam described by Durnin's solution has no radial boundaries, and therefore it possesses the infinite energy. However, it was immediately noticed and implemented experimentally (Durnin et al., 1987-2) that the spatial spectrum of an ideal Bessel beam in k-space lies on a ring, which is single off-axis delta function. Therefore, the simplest way to generate space-limited Bessel beam is to put the collimated input beam through an annular aperture placed in the back focal plane of a lens. However, this approach is energetically inefficient. Now the more popular methods for generating quasi-diffraction-free Bessel beam are employing an axicon (conical lens) or spatial light modulator (SLM). The Bessel beam generated by these methods creates the narrow central peak with a length significantly exceeding the Rayleigh length of tightly focussed Gaussian beam. We describe in details the applications of the Bessel-like beam in micro-machining and for generation of extreme energy density in Chapter 6.

1.2 The Matter

We are going to consider the intense light beam interaction with the different solids, metals and dielectrics. The material and optical properties of solids are changing in the process of such an interaction affecting in turn the electric field inside a solid. Therefore, in general the interaction should be described by the coupled set of the Maxwell and material equations accounting for the change in the space/time of the material properties. The most dramatic changes occur in the intense laser-affected dielectrics. At low intensity, the number of conduction electrons in the valence band is small; therefore, the imaginary part of the dielectric permittivity is close to zero. That is the reason for the wide-band gap dielectrics transparency at low intensity. The light absorbs mainly on the conduction electrons (in the presence of the third body, phonon or ion to ensure the momentum conservation). At the increasing intensity, the ionization of a dielectric begins. The number of electrons in conduction band increases rising the absorption. The laser-affected permittivity becomes time and space dependent. The interaction with metals and dielectrics is essentially different and will be considered in details in the other chapters. However, the contribution of the conductivity electrons into the laser-affected permittivity in many cases is dominant. Thus, it is instructive to consider the simple model of the high-frequency electric field interaction with the free electrons and to introduce some major notions.

1.2.1 Electrons' Oscillations and Scattering in the High-Frequency Electromagnetic Field

We are considering laser–matter interaction at the non-relativistic intensity. It means that velocity of the field-accelerated electron is much less than the speed of light in vacuum. Therefore, the Lorentz force of the magnetic field can be omitted (because it is proportional to the ratio of electron velocity to the speed of light in vacuum). The free electron can only scatter a photon but cannot absorb it because in this case it is impossible to comply with the conservation laws. Thus, three body interactions involving a photon, an electron and an atom (or ion) are responsible for the absorption of laser energy. In the absence of the external field, this spontaneous process is well known as bremsstrahlung emission. In the presence of external field the two processes occur: the induced photon emission and inverse process of photon absorption. The net result is the photon absorption process called inverse bremsstrahlung or collisional absorption. All collisions of electrons with electrons, ions and neutrals should be taken into account because the electron–electron collisions only do not change the total electron momentum and the total electron current.

Let us consider the free electron oscillations in the laser field, $\bar{E} = E_0 \cdot e^{-i\omega\tau}$ assuming that the ions make a neutralizing background. The oscillation energy can be converted into the electron's kinetic energy by the multiple electron collisions with the lattice (phonons) or with ions. The Newton equation for the electron's motion in the high-frequency linearly polarized electric field is as follows:

$$\frac{dv_e}{dt} = \frac{e \cdot \bar{E}}{m} - \nu_{\text{eff}} \cdot v_e \tag{1.9}$$

The energy of the electric field converts to the electron's energy via the collisions with the effective rate, ν_{eff}. The solution to this equation in the form $v_e = v_0 \cdot e^{-i\omega\tau}$ is straightforward:

$$v_e = \frac{e \cdot \bar{E}}{m \cdot (\nu_{\text{eff}} - i\omega)} \tag{1.10}$$

when absorption is low, $\nu_{\text{eff}} << \omega$, the electrons simply oscillate in the high-frequency electric field with the amplitude velocity $\bar{v}_{\text{osc}} = e\bar{E}/m\omega$. The energy of oscillations averaged over the many wave periods is a convenient characteristic of the applied field strength. We present the oscillation energy in the general form applicable for the circular ($\alpha = \pm 1$) and linear ($\alpha = 0$, as above) polarization:

$$\varepsilon_{\text{osc}} = \frac{e^2 \cdot E^2 \cdot (1 + \alpha^2)}{4m \cdot \omega^2} \tag{1.11}$$

We present the above formula in the form convenient for numerical estimates as follows:

$$\varepsilon_{\text{osc}}(\text{eV}) = 9.375 \cdot (1 + \alpha^2)(I/10^{14}\,\text{W/cm}^2) \cdot (\lambda/\mu\mu)^2 \tag{1.12}$$

For example, the electron's oscillation energy in the field of linear polarized laser beam at wavelength of one micron at average intensity 10^{14} W/cm^2 equals to 9.375 eV close to the ionization potential for many atoms. The oscillation velocity of electron in such a field is $v_{\text{osc}} \approx 2 \times 10^8$ cm/s, which is close to the inter-atomic velocity. Thus, the interaction of laser at such intensity converts practically any material to the plasma state.

1.2.2 The Drude Model for the Permittivity of Simple Plasma in the High-Frequency Electric Field

Above we considered the individual electron motion. Let us now introduce the electrons' current density, $j = e \cdot n_e \cdot v_e$, taking into account the collective behaviour of the electrons. Using (1.10) and comparing with Ohm's law $j = \sigma \cdot \bar{E}$, one gets the conductivity of such plasma:

$$\sigma = \frac{e^2 \cdot n_e}{m(v_{\text{eff}} - i\omega)} \tag{1.13}$$

The relation between the dielectric function and the conductivity for simple electronic plasma (number of electrons equals to that of immobile ions) is the following:

$$\varepsilon_{\text{el}} = 1 + i\frac{4\pi\sigma}{\omega} \tag{1.14}$$

The subscript "el" denotes that this part of the dielectric permittivity relates to the conductivity of electrons in metals (or in simple electronic plasma). Now the dielectric permittivity attains well-known form of the Drude-like function (Landau et al., 1984):

$$\varepsilon_{\text{el}} = 1 - \frac{4\pi e^2 \cdot n_e}{m_e\omega(iv_{\text{eff}} + \omega)} \tag{1.15}$$

One can see that the permittivity includes the important characteristic of collective plasma behaviour, the electron plasma frequency, ω_{pe}:

$$\omega_{\text{pe}}^2 = \frac{4\pi \cdot e^2 n_e}{m_e} \tag{1.16}$$

Now the permittivity takes a form

$$\varepsilon_{\text{el}} = 1 - \frac{\omega_{\text{pe}}^2}{\omega^2(1 + iv_{\text{eff}}/\omega)}. \tag{1.17}$$

Two important characteristics of the laser plasma interactions are introduced in this permittivity form: plasma frequency explicitly and critical electron density implicitly. Let us discuss both notions explicitly.

1.2.2.1 Electrons' plasma frequency

The electrons' plasma frequency, frequency of charge density oscillations, is the fundamental characteristic of quasi-neutral plasma. Let us discuss it in more details.

In quasi-neutral plasma where the number of negative charges (electrons) equals to the number of positive charges (ions) particles cannot be in a rest even at zero temperature due the Earnshaw theorem (consequence of the Gauss law). It follows from the Gauss law that in the equal mixture of positive and negative charges the net electric force always acts at any charge in the plasma. Thus, infinitesimal charge density fluctuations inevitably appear. Electrons perform oscillations due to the action of the electrostatic field induced by the electric charge fluctuations with the so-called plasma frequency depending on the electrons number density, $\omega_{\text{pe}}^2 = \frac{4\pi e^2 n_e}{m_e}$. The same electrostatic field acts on the ions, which oscillate with much slower ion's plasma frequency, $\omega_{\text{ie}}^2 = \omega_{\text{pe}}^2 \frac{m_e}{M}$. Here, it is suggested that the number density of ions equals to the electrons number density. Due to huge differences in masses, the motion of ions in many cases could be neglected. Ions serve as a positive background ensuring the charge conservation (see Appendix 2).

1.2.2.2 Critical density of electrons

Let us present the permittivity of (1.17) as a sum of real and imaginary part as follows:

$$\varepsilon_{\text{el}} = 1 - \frac{\omega_{\text{pe}}^2}{\omega^2\left(1 + \nu_{\text{eff}}^2/\omega^2\right)} + i\frac{\omega_{\text{pe}}^2}{\omega^2\left(1 + \nu_{\text{eff}}^2/\omega^2\right)}\frac{\nu_{\text{eff}}}{\omega} \tag{1.18}$$

The effective collision rate dramatically changes during the interaction time in a different way in metals and dielectrics strongly affecting the interaction process. It will be considered in details later in the book.

The number density of the conductivity electrons in the dielectric under the action of the intense laser continuously increases due to ionization while the absorption is still negligible, $\nu_{\text{eff}} << \omega$. The real part of the permittivity at this stage reads $\varepsilon_{\text{el}} \approx 1 - \omega_{\text{pe}}^2/\omega^2$. The number density of the generated electrons gradually reaches the level where the plasma frequency equals to the laser frequency and the real part of the permittivity turns to zero. This is a critical point in the interaction process when the running electromagnetic wave transforms to the evanescent wave. The critical electron density then is derived from the condition of equality of the laser frequency to the plasma frequency, $\omega = \omega_{pe}$, in the form

$$n_c = \frac{m_e \cdot \omega^2}{4\pi \cdot e^2} = 1.116 \cdot 10^{21} \left(\frac{\mu\text{m}}{\lambda} \right)^2 [\text{cm}^{-3}]. \tag{1.19}$$

1.2.3 Absorbed Energy Density

The absorbed energy density for the plane wave normally incident along the x-direction on the plane surface (y, z) can be calculated in a rather general form not specifying the mechanism of absorption. Indeed the energy of electromagnetic field absorbed per unit time per unit volume expresses through the divergence of the Poynting vector as follows:

$$Q_{\text{abs}} = -\nabla \cdot S = -\nabla \cdot \left(\frac{c}{4\pi} E \times H \right) \tag{1.20}$$

Replacing the space derivatives through the time derivatives from the Maxwell equations and then averaging by the many light periods (slow envelope approximation) results in the familiar form:

$$Q_{\text{abs}} = \frac{\omega}{8\pi} \varepsilon_{\text{im}} \, |E_a|^2 \tag{1.21}$$

Here E_a is the electric field inside a solid, ε_{im} is the imaginary part of the dielectric permittivity. Now we should relate the field inside a solid to the laser field at the normal incidence by the boundary conditions of continuity E_y and H_z as follows:

$$\lfloor E_a(0) \rfloor^2 = \frac{4}{\lfloor 1 + \varepsilon^{1/2} \rfloor^2} \lfloor E_0 \rfloor^2 \tag{1.22}$$

Now formula (1.21) takes a general form appropriate for estimation of the absorbed energy density in any medium:

$$\langle Q_{\text{abs}} \rangle = Re \frac{1}{2} Q_{\text{abs}} = \frac{\omega \varepsilon_{\text{im}}}{8\pi} \frac{4 \lfloor E_0 \rfloor^2}{\lfloor 1 + \varepsilon^{1/2} \rfloor^2} \left[\text{W/cm}^3 \right] \tag{1.23}$$

Let us transform the above formula to the form revealing the physical meaning of all terms. It is useful introducing the refractive index as, $\varepsilon_{\rm el} = (n + ik)^2$. One gets $\varepsilon_{\rm im} = 2nk$, while the Fresnel absorption coefficient is expressed as $A = \frac{4n}{\lfloor 1+\varepsilon^{1/2} \rfloor^2}$. Introducing the incident intensity, $I_0 = c|E_0|^2/8\pi$, and the absorption length, $l_s = c/\omega k$, the formula (1.23) transforms into the following:

$$Q_{\rm abs} = \frac{2A}{l_s} I_0(t) \ [\mathrm{W/cm^3}] \tag{1.24}$$

Integrating (1.24) by time under assumption that absorption coefficient and length remain constant one obtains the total energy absorbed during the time t:

$$E_{\rm abs} = \int_0^t Q_{\rm abs} d\tau = \frac{2A \cdot F(t)}{l_s} \left[\frac{\mathrm{Joules}}{\mathrm{cm^3}} \right]; \ F(t) = \int_0^t I_0(\tau) d\tau \tag{1.25}$$

Here $F(t)$ is the energy incident per unit area (fluence) during the time.

1.2.4 Hierarchy of the Laser-Affected Material Transformations as a Function of Laser Intensity/Fluence

With the help of formula (1.25), one can estimate the fluence and average intensity necessary for producing different material modification with a laser of known pulse duration and wavelength. As a guide, one can use the energy density necessary for material modification in equilibrium. However, we shall indicate that the state of short intense laser-affected material during the pulse and long after the end of the pulse remains different from the familiar equilibrium transformations like melting and evaporation, which we describe later in the book. The general picture for the material modification from excitation of phonons at low intensity/fluence up to the swift evaporation (coined by ablation in non-equilibrium) is presented in Fig. 1.4. As examples of such calculation, we present estimates for the energy per pulse, intensity and fluence necessary for melting and ablation of metals.

1.2.4.1 Melting

The enthalpy of melting a metal in equilibrium (when electrons and atoms have the same temperature) is in the order of kJ/cm^3. Absorption depth in metals is of the order of several tens of nanometres (1 nm = 10^{-7} cm), $l_s \approx 10^{-5}$ cm. Thus, in order to deposit the energy density of kJ/cm^3 by, 100 fs, $\lambda \sim 1$ μm, laser one needs absorbed fluence of 10^{-2} J/cm^2 and intensity $\sim 10^{11}$ W/cm^2.

1.2.4.2 Ablation

The enthalpy of evaporation in equilibrium for good metals is around (40–50) kJ/cm^3. In order to achieve such energy density with the laser (100 fs, $\lambda \sim 1$ μm) one needs in accord with (1.25) the average intensity of (4–5) × 10^{12} W/cm^2 and absorbed fluence of (0.4–0.5) J/cm^2. The energy expenses for modification of dielectrics are significantly larger due to necessity of ionization for achieving a substantial absorption. We discuss the processes of modification of metals and dielectrics in more details in the other chapters.

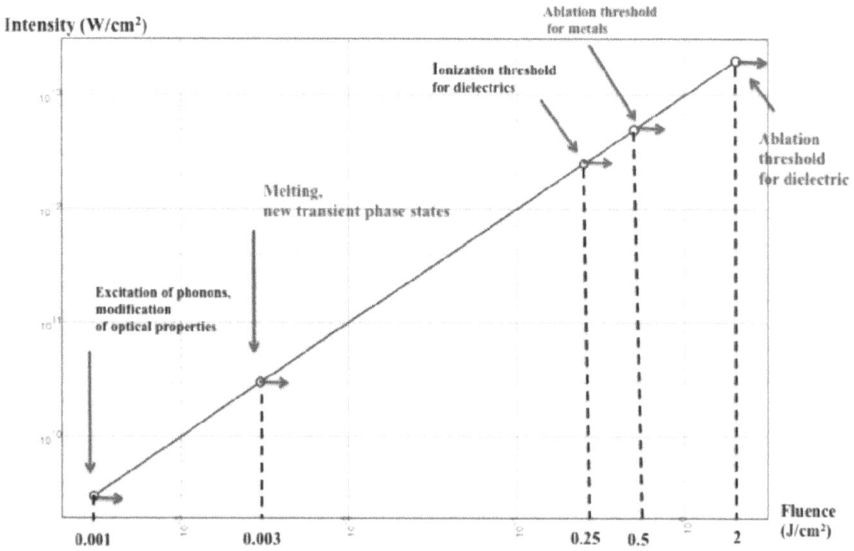

Figure 1.4 Material modifications vs. intensity/fluence performed by 100 fs pulse at 800 nm.

It is instructive to express the further milestones for the intensity increase well over the ablation threshold through the fundamental constants: electronic charge, *e*, free electron mass, m_e, and Plank constant, $\hbar$.

1.2.4.3 Atomic field intensity

The next milestone in the intensity rise over the ablation threshold is achieved when the field amplitude in the laser beam equals to the inter-atomic electric field, which is the electric field created by electron at the first Bohr orbital:

$$E_a = e/a_B^2 = 5.76 \times 10^9 \text{ V/cm}$$

The space scale is the Bohr radius:

$$a_B = \hbar^2/m_e e^2 = 5 \times 10^{-9} \text{ cm}$$

Note that the work produced by the atomic field on electron at the distance of the Bohr radius is equal to the ionization potential of hydrogen, $eE_a a_B = J_0 = 13.6 \text{eV}$. The corresponding *atomic intensity* for linearly polarized electric field, averaged by the many light periods, reads

$$I_a = \frac{cE_a^2}{8\pi} = \frac{c}{8\pi}\frac{e^2}{a_B^4} = \frac{\hbar c}{8\pi}\frac{\omega_a}{a_B^3}. \tag{1.26}$$

The fluence corresponding to the inter-atomic intensity appears to be independent on the atomic time:

$$F_a = I_a \times \omega_a^{-1} = \frac{\hbar c}{4\pi a_B^3} = 2 \text{ Joule/cm}^2$$

The above fluence is close to the ablation threshold for majority of dielectrics, that is the laser pulse duration independent, as it was demonstrated experimentally and understood theoretically. This is described in Chapter 5. The energy scale, the Rydberg (doubled ionization potential, J_0) and the frequency scale, *atomic* frequency, ω_{at}, are as follows:

$$2J_0 = e^4 m_e / \hbar^2 \approx 27.2 \text{ eV}$$
$$\omega_{at} = J_0/\hbar^2 = 2.066 \cdot 10^{16} \text{ s}^{-1}$$

Note that the characteristic atomic velocity scale is $v_a = \frac{e^2}{\hbar} \approx 2.2 \times 10^8$ cm/s. Typical *atomic time scale* is defined as $t_a = \frac{a_B}{v_a} = \hbar/J_0 \approx 50$ attoseconds. Thus, one can separate the interaction regimes by the natural milestones emerging along with the intensity increase. Any material by the action of laser at the atomic intensity is converted to plasma as we will see in more details later.

1.2.4.4 The relativistic intensity

The relativistic effects become essential in the description of the interaction process when the oscillation energy of electron in the high-frequency field becomes comparable to the rest energy of electron, $\varepsilon_{\text{rest}} = m_e c^2 \approx 500\, k$ eV. At $I \geq I_{\text{rel}}$, physics of interaction is changed. The electric force and the Lorentz force become comparable. The relativistic intensity follows from the matching the rest energy to the non-relativistic oscillation energy, $\frac{e^2E^2}{4m_e\omega^2} = m_ec^2$. For the plane polarized field one gets

$$I_{\text{rel}} = \frac{m_e^2\omega^2c^3}{2\pi e^2} = 5.48 \cdot 10^{18} \left(\frac{\mu m}{\lambda}\right)^2 \text{W/cm}^2. \tag{1.27}$$

Thus, for $\lambda = 1$ μm the relativistic intensity equals to 5.48×10^{18} W/cm^2. We consider later in the book the laser/matter interactions with the wavelength in the middle of visible range $\lambda \leq 1$ μm and the laser intensity lower than the relativistic intensity.

1.3 Summary

- Intensity averaged over many light periods $I = \frac{cE^2}{8\pi}(\alpha^2 + 1)$ (circular polarization, $\alpha = \pm 1$; linear polarization, $\alpha = 0$).
- The focal spot radius, $r_f = 0.61\lambda/\text{NA}$; $\text{NA} = n \times \sin\theta$

- The energy of the free electron's oscillations in the high-frequency field

$$\varepsilon_{\rm osc}({\rm eV}) = 9.375\cdot(1+\alpha^2)(I/10^{14}\,{\rm W/cm^2})\cdot(\lambda/\mu m)^2$$

- The frequency of charge density oscillations in quasi-neutral plasma

$$\text{Electrons' plasma frequency, } \omega_{\rm pe}^2 = \frac{4\pi\cdot e^2 n_e}{m_e};$$

$$\text{Ion's plasma frequency, } \omega_{\rm ie}^2 = \omega_{\rm pe}^2\frac{m_e}{M}$$

- Critical density of electrons ($\varepsilon_{\rm re} = 0$; $\omega = \omega_{\rm pe}$)

$$n_c = \frac{m_e\cdot\omega^2}{4\pi\cdot e^2} = 1.116\cdot 10^{21}\left(\frac{\mu m}{\lambda}\right)^2\ [{\rm cm^{-3}}]$$

- Absorbed energy density rate, $Q_{\rm abs} = \dfrac{2A}{l_s}I_0(t)\ [{\rm Joules/cm^3\cdot s}]$
- Hierarchy of the laser-induced material transformations considered in the book along the maximum intensity during 100 fs-long pulse:
  - from the reversible transformations at the intensity below 10^{12} W/cm^2
  - to the extreme energy density exceeding strength of any existing materials at intensity $\geq 10^{14}$ W/cm^2.

Chapter 2

Interaction with Metals

The material and optical properties of solids under the intense light beam action are changing in the process of interaction affecting in turn the electric field inside a solid. Ideally, the interaction should be described by the coupled set of the Maxwell and material equations accounting for the modification of the material properties interconnected with the field changes:

$$\nabla \times \nabla \times E = -\frac{1}{c^2}\frac{\partial^2 D}{\partial t^2}$$
$$D = \varepsilon E$$
$$div\, D = 0$$
$$\varepsilon = \varepsilon_{\text{val}} + \varepsilon_{\text{cond}} \tag{2.1}$$

The dielectric permittivity in general includes the contribution of valence and conduction band electrons. Both contributions are important in the intense laser-affected dielectrics. However, the permittivity in metals is well described in the Drude approximation when the contribution of conduction electrons is dominant.

Femtosecond Laser-Matter Interactions: Solid-Plasma-Solid Transformations at the Extreme Energy Density
Eugene Gamaly

ISBN 978-981-4877-40-4 (Hardcover), 978-1-003-25661-8 (eBook)
www.jennystanford.com

The number density of the conductivity electrons can be considered constant during the interaction time shorter of the major relaxation processes and with a fluence range up to the ablation threshold, It is not the case for the semi-metals. Indeed, in Bi, the number density of the electrons in the conduction band increases even during the melting as we show later in the book. However, even for good metals such as gold and silver, where *d*-electrons are close to the Fermi surface, the ablation thresholds are well described under assumption of the unchanged number density of the conduction electrons (see Chapter 5). Probably, in this case the laser intensity (in a range of a few TW/cm^2) is not high enough to affect the state of *d*-electrons.

The major changes of the permittivity in the laser-affected metals are related to the modification of the effective collision rate of the electrons along with the increased intensity of the laser. The effect on the optical properties relates to the changes in the momentum transfer rate from electron to the rest of the medium. The energy transfer is controlled by the energy transfer rate. The energy distribution of the conduction electrons at room temperature is described by the Fermi–Dirac function. During the fast heating by the short intense pulse, the distribution function approaches the Maxwell distribution at $k_B T_e \geq \varepsilon_F$ (ε_F is the Fermi energy in the electrons energy distribution) in two-step way as we show later in this chapter. Along with the change in the energy distribution, the heat capacity of electrons and ions is also modified. All these modifications are considered in this chapter.

2.1 Interaction of the Plane Wave with a Metal Layer

Let us consider the simplified interaction case where one can see the main features of interaction. We approximate the laser beam by the plane linearly polarized wave incident along the normal to the surface of semi-infinite metal in the *x*-direction from the vacuum (Fig. 2.1). The electric field is described as follows: $\bar{E} = \bar{E}_0 \cdot \exp(i\omega t)$; $\bar{E}(0, E_y, 0)$. Let us consider the case when the complex dielectric permittivity of a metal, ($\varepsilon = \varepsilon_{re} + i\varepsilon_{im}$), is not affected during the interaction, i.e. ε = const. at $x > 0$. Thus, the second equation in (2.1), $\nabla D = \nabla \cdot \varepsilon E = 0$, reduces to $\nabla \cdot E = 0$. The Maxwell Eq. (2.1) takes the form of one-dimensional Helmholtz equation:

$$\frac{d^2 E_y}{dx^2} + \frac{\varepsilon \omega^2}{c^2} E_y = 0 \tag{2.2}$$

The solution is well known (Landau and Lifshitz, 1984):

$$E = E_0 \exp(ik_0 \cdot \varepsilon^{1/2}) = E_0 \exp(ik_0 nx - x/l_s);\ k_0 \equiv \frac{\omega}{c} \tag{2.3}$$

The wave, penetrating inside a metal, decays exponentially with the length of e-fold decrease in the amplitude (skin-depth or absorption depth) depending on the imaginary part of the refractive index, $\sqrt{\varepsilon} = n + ik$ as follows:

$$l_s = \frac{c}{\omega \cdot \kappa} \tag{2.4}$$

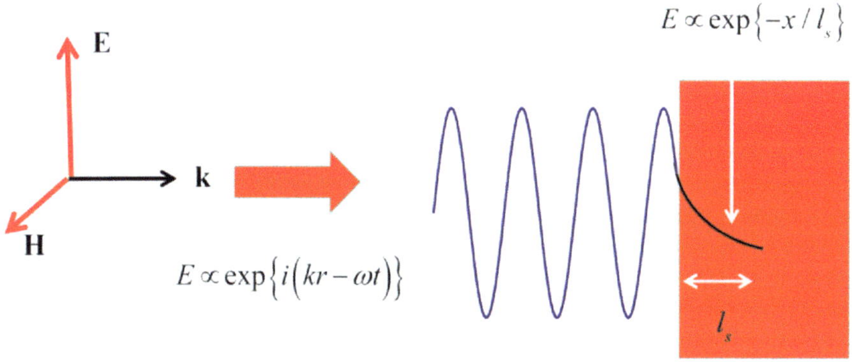

Figure 2.1 Plane electromagnetic wave penetration inside a metal.

The optical properties of the laser-excited metal, ε and κ, are functions of the laser intensity through the electron–phonon momentum exchange rate. Therefore, the characteristic length scale varies from several tens of nanometres (nm) at high intensity to a few hundred of nanometres at low intensity according to (2.4).

This interaction characterizes by three scale lengths: electrons mean free path, l_{ee}, the skin depth, l_s, and the scale length controlling the time dispersion (application of Ohm's law), v_F/ω. The above solution is the case so-called normal skin effect when the relation between three characteristic lengths should obey the following conditions: $l_{ee} << l_s$; $l_{ee} << v_F/\omega$. The first condition fulfils in the majority of the interactions of interest. However, the second condition might be violated at low laser frequency.

Let us consider as an example the interaction with the good metal (aluminium) having three conduction electrons per atom (electron number density $n_e = 1.806 \times 10^{23}$ cm^{-3}; corresponding plasma frequency, $\omega_{pe} = 2.39 \times 10^{16}$ s^{-1}; Fermi energy $\varepsilon_F = 11.63$ eV; and Fermi velocity, $v_F = 2.02 \times 10^8$ cm/s). At low intensity and at the room temperature, the absorption of laser light with wavelength in a visible range is low. For example, in Al at $\lambda = 532$ nm, the absorption constitutes only 0.075 nm (see Appendix 5).

The hydrodynamic motion begins after the energy transfer from electrons to the lattice. Time for the energy transfer to the lattice is in the picosecond range as we show later. Thus, during the ultra-short pulse, the hydrodynamics is negligible and the vacuum-solid boundary maintains a perfect density step. In this case, the interaction falls in the frame of the normal skin effect at the non-relativistic intensity.

The major mechanism of absorption is the inverse bremsstrahlung absorption, which accounts for electron collisions with ions. In the quantum mechanical language, this process is described as the absorption of a photon by an electron in the presence of the third body (atom or ion). Classically, this process is described as the energy gain by the electron oscillating in high-frequency external field through collisions. Therefore, the increase in absorption strongly depends on the collision rate, which we consider later in the chapter.

Conventionally two polarization states are considered: in *s*-polarized beam the electric field is perpendicular to the plane of the incidence, while at p-polarization it lies in the plane of the incidence. At normal incidence, s and p polarization states are identical.

2.2 Electron and Lattice Temperature, Energy Balance, Two-Temperature Approximation (Landau et al., 1984; Kaganov et al., 1957; Anisimov et al., 1974)

The change in the electrons energy per unit time during the period shorter than the energy transfer time equals the absorbed energy rate from (1.24) minus losses to energy transfer to lattice (ions) and for the energy transport out of the focal spot by the electronic heat conduction:

$$\frac{dE_e}{dt} = Q_{\text{abs}} - Q_{\text{ei}} - Q_{\text{heat}} \tag{2.5}$$

Correspondingly, the energy delivered to the lattice (ions) reads

$$\frac{dE_L}{dt} = Q_{\text{ei}}. \tag{2.6}$$

Here we conventionally assume, that the average (over the energy distribution) energy in each of species is the electrons' and lattice "temperature", T_e, T_L (two-temperature approximation). Then the electrons' and lattice energy densities are as follows:

$$E_e = C_e \cdot n_e \cdot T_e; \quad E_L = C_L \cdot n_a \cdot T_L \tag{2.7}$$

Here C_e, C_L are the electronic and lattice heat capacity, which are changing during the intense pulse action.

The absorbed energy is contained solely in the electron subsystem at the end of the ultra-short pulse. The electrons in metal with the number density n_e at the low temperature, $k_B T_e << \varepsilon_F$, fill the energy levels up to the Fermi energy $\varepsilon_F(n_e)$. The electrons are treated as the degenerated electron gas. The heat capacity of the degenerated gas at low temperature reads (Landau et al., 1984):

$$C_e \approx \pi^2 k_B^2 T_e / 2\varepsilon_F \tag{2.8}$$

The number density of the conductivity electrons is practically unchanged up to the ablation threshold. The heat conduction losses and ions heating are negligible. Therefore, the internal energy of the electrons in a metal at any time during the pulse equals the absorbed energy density to this moment. The electrons' temperature reaches the maximum at the end of the pulse. It can be retrieved from the equality of the absorbed energy to the internal energy of electrons making use of Eq. (1.25), as follows: $E_e = E_{\text{abs}}; \rightarrow C_e \cdot n_e \cdot T_e = \frac{2A \cdot F(t_p)}{l_s}$. The electron temperature to the end of the pulse follows from the equation:

$$k_B T_e = \frac{2A \cdot F(t_p)}{l_s \cdot C_e(T_e) \cdot n_e} \tag{2.9}$$

Thus, the average electron energy grows in direct proportion to increasing absorbed fluence. The energy conservation law allows estimating the temperature after the temperature equilibration, $T = T_e = T_i$. Indeed, the sum of the electrons and atoms internal energy equals the absorbed energy density as follows:

$$k_B T = \frac{2A \cdot F(t_p)}{l_s \{C_e(T) \cdot n_e + C_L(T) \cdot n_a\}} \tag{2.10}$$

The temperature by (2.10) should be calculated with the heat capacities corresponding to the electrons and lattice states at the moment of equilibration.

2.3 Temperature Dependence of the Electronic and Lattice Heat Capacity

With the growing electron temperature, electrons are gradually converted from a degenerate state to the electronic plasma. As it follows from (2.8) at $k_B T_e = \frac{3}{\pi^2} \varepsilon_F$ the electronic heat capacity of (2.8) is increased to that of the ideal gas with the heat capacity $3k_B/2$ (see Fig. 2.2). In reality it occurs at lower (not known exactly) temperature.

The atoms at low temperature in a solid have six degrees of freedom. Accordingly, the atomic specific heat is equal to $3k_B$ per atom. The vibrational motion of atoms changes to purely translation mode when the temperature gradually increases. Correspondingly the atomic specific heat decreases to $3k_B/2$ per atom. The potential barrier against the free motion of atoms through the solid is naturally related to the binding energy, ε_b. The effective temperature, where transition to free atoms specific heat occurs, was established as, $k_BT_b = 2\varepsilon_B/3$ (Zel'dovich and Raizer, 2002). Thus, the increase in the lattice temperature up to several electron-volts is accompanied by an increase in the specific heat of electrons and decrease in the heat capacity of atoms. Both electron and ion heat capacities are aiming to same value as for the ideal gas. The qualitative dependence of electrons' and ions' heat capacity is presented in Fig. 2.2.

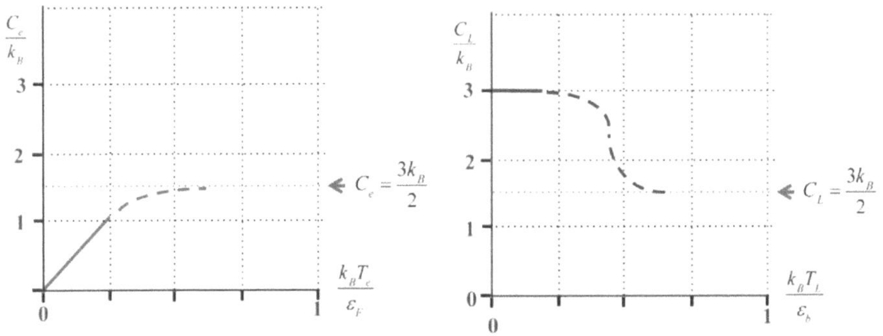

Figure 2.2 Qualitative dependence of the electron (C_e) and lattice (C_L) specific heat on the relative electron and lattice temperature, respectively.

The binding energy in metals is in a range of 3–5 eV. Therefore, transition to the ideal gas heat capacity occurs at 2–3 eV. The Fermi energy in metals is in a range 5–14 eV. Hence, the conversion of the electronic specific heat happens in a close range of 2–5 eV.

One might consider the states of an excited metal at the temperature below 1 eV as a solid phase while those above 10 eV as a plasma state.

2.4 Electrons Relaxation Processes in the Laser-Affected Metal

The electrons convert part of the energy of their oscillation under the imposed laser field into the internal energy by collisions. The fastest (fraction of femtosecond) are the electron–electron collision establishing the local energy distribution in the electron sub-system. The electrons establish the conventional Fermi–Dirac distribution at low excitation and low temperature, $k_B T_e << \varepsilon_F$. At high excitation, $k_B T_e \approx \varepsilon_F$, the electron distribution function is reconstructed to the form close to that of the Maxwell–Boltzmann. On the time scale longer than that for electron–electron collisions electrons start exchanging momentum with a lattice (ions). The momentum exchange rate is responsible for the light absorption and for the electronic heat conduction.

The energy transfer from hot electrons to the cold lattice occurs on a time scale of a few picoseconds increasing the lattice temperature and building the atomic energy distribution. At the time when the electrons and lattice temperature equilibrate the atomic distribution has the form of the incomplete Maxwell distribution lacking the high-energy tail. The creation of particles with the energy much larger than the average needs much more collisions and therefore the longer time for the tail to be filled in. The time for building up the high-energy tail in the ions distribution function appears to be the longest of all relaxation times. We begin with the fastest of all relaxation processes of electron–electron collisions.

2.4.1 Electron–Electron Collisions

The collision rate or probability for collision per unit time in the kinetic theory conventionally expresses through the number density of colliding particles, n_e, cross section for the interaction, σ_{ee}, and relative velocity of colliding particles, v_e, as follows:

$\nu_{ee} = n_e\sigma_e v_e$. The cross section for the electron–electron Coulomb collision rate is inversely proportional to the squared electron energy.

However, the studies of electron collisions for the small excitations over the Fermi level (Quinn and Ferrell, 1958; Pines, 1964) demonstrated that electron scattering rate grows in proportion to the plasma frequency and the electron energy reaching the maximum at the Fermi energy as follows:

$$\nu_{ee} = \frac{v_F}{l_{ee}} = \frac{\omega_{pe}}{1.88} \cdot \frac{\varepsilon_e}{\varepsilon_F} \tag{2.11}$$

Plasma frequency for good metals ($n_e \approx 10^{23}$ cm^{-3}) is around $\omega_{pe} \approx 2 \times 10^{16}$ s^{-1}, while $\varepsilon_F \approx 10$ eV. Therefore, for the electrons, heated to the energy around electron-volt, the electron–electron collision rate is $\sim 10^{15}$ s^{-1}. Thus, the time necessary for establishing the main part of the electronic distribution is around femtosecond.

2.4.2 Electron–Phonon Momentum Transfer

Electrons-to-lattice coupling

We are considering the electrons heating during the time much shorter than the period of the harmonic vibration of the lattice, phonons. On the quantum language the electron–phonon interaction is emission/absorption of phonons with the total energy and quasi-momentum conserved. On the other hand, the electron-to-lattice coupling may be considered as a distortion of a lattice. The electrons are emitting phonons (distorting a lattice) when the electrons' group velocity exceeds the acoustic velocity (Kittel, 1996). Two approaches are compatible and applicable for the estimates the coupling in non-equilibrium. Let us first demonstrate that the electrons-to-lattice coupling always exists in a solid and may be expressed through the fundamental constants.

The kinetic energy of atoms experiencing the harmonic vibrations around the equilibrium positions equals the potential energy. The last one expresses through the quadratic term for expansion of the inter-atomic potential in the powers of atomic displacements, Δ_{ph}, as follows (Il'inskii and Keldysh, 1994):

$$\frac{p_{\text{ph}}^2}{2M} \approx \frac{1}{2}\frac{\partial^2 U}{\partial r^2}\Delta_{\text{ph}}^2 \tag{2.12}$$

The uncertainty relation, $p_{\text{ph}} \cdot \Delta_{\text{ph}} \geq \hbar/2$, connects the momentum and amplitude of phonons considered as the quantum quasi-particles. We estimate, $\frac{\partial^2 U}{\partial r^2} \approx \frac{J_0}{a_B^2}$, through the fundamental constants, ionization potential $\left(J_0 = \frac{e^2}{2a_B}\right)$ and the Bohr radius (see Chapter 1). Thus, the coupling of the electrons with a lattice was implicitly introduced. Now, the phonon's amplitude (lattice distortion) relates to the Bohr radius through the ratio of electron's to atomic mass, $\Delta_{\text{ph}} \approx \left(\frac{m_e}{M}\right)^{1/4} \cdot a_B$ (adiabatic principle). Expression for the phonons energy $\hbar\omega_{\text{ph}} \approx M\Delta_{\text{ph}}^2\omega_{\text{ph}}^2$ allows getting the phonon's frequency through the phonon's amplitude, $\omega_{\text{ph}} \approx \hbar / M\Delta_{\text{ph}}^2$. In non-equilibrium, we may consider the phonon frequency as a rate of a lattice distortion (dilation). Using the above formulae for the ionization potential, and for the phonons amplitude one obtains the expression for the phonon's frequency through the atomic frequency, $\omega_a \approx J_0/\hbar$, as follows:

$$\omega_{\text{ph}} \approx \left(\frac{m_e}{M}\right)^{1/2} \cdot \omega_a \tag{2.13}$$

This is the general relation for the electrons-to-lattice coupling but lacking the explicit dependence on the electron energy, which we present below.

The electron–phonon collision rate dependence on the electron energy

During the ultra-short pulse action on metal, the electrons' energy increases along with the growing intensity while the number density of electrons and atoms, atomic positions and the lattice temperature remains practically unchanged. Therefore, the electronic mean free path between collisions, $l_{\text{mfp}} = (n\sigma)^{-1} \approx d_a$ remains constant and approximately equal to the nearest-neighbour distance. Hence, the electron-to-lattice collision rate grows up as the electron energy increases

$$\nu_{e-L} \approx v_e/l_{mfp} \propto (k_B T_e)^{1/2} \quad (2.14)$$

The electron–phonon collision rate grows up as the electron energy increases up to the natural maximum determined by the nearest-neighbour distance and the electrons velocity, $\nu_{e-ph}^{max} \approx v_e^{max}/d_a$ (Allen, 2000). It occurs when the electrons' energy approaches to the Fermi energy, $k_B T_e \approx \varepsilon_F$. A material at such temperature is converted to plasma. The electron–ion collision rate in plasma decreases as the electron temperature grows up, $\nu_{e-i} \propto T_e^{-3/2}$ (see later in the chapter). The theory of the transition from the electron–phonon momentum exchange in solid to electron–ion Coulomb collisions in plasma does not exist to the best of the author's knowledge. The reasonable interpolation describing the collision rate around the maximum for the numerical calculations of the ultra-short pulse interaction with metals was proposed by Eidmann et al. (2000).

Hence, the momentum exchange rate in the interaction of the ultra-short pulses with good metals from the room temperature ($k_B T_{room}$ = 0.025 eV) up to the maximum can be estimated by the following formula:

$$\nu_{e\text{-ph}}^{mom} = \nu_{e\text{-ph}}^{mom}(T_{room}) \cdot \left(\frac{k_B T_e}{0.025\ \text{eV}} \right)^{1/2} \quad (2.15)$$

The transient electron-to-lattice momentum exchange rate affects the dielectric permittivity of the laser-excited metal. Therefore, it is appropriate to use the collision rate extracted from the optical measurements at the room temperature in the scaling above. For example, the value of the collision rate for Al from the reflectivity measurements, $\nu_{e\text{-ph}}(T_{room}) = 8.5 \times 10^{14}\ \text{s}^{-1}$ has been extracted in the reference (Eidmann et al., 2000). It is easy to see that the above scaling gives at $k_B T_{room}$ = 10 eV the maximum value close to the estimated maximum.

2.4.3 Electron–Phonon Energy Transfer

The electron-to-lattice energy transfer leads to the electrons and lattice temperature equilibration usually after the end of the laser pulse. Conventionally the electrons-to-lattice coupling considered as emission and absorption of phonons. The energy of emitted phonon is the energy transferred by an electron to the lattice.

The ultra-short laser–matter interaction time is typically shorter than the period of harmonic vibrations of a lattice, phonons. Hence, it is more appropriate considering the electrons-lattice coupling as a distortion of the lattice (Kittel, 1996). The absorbed part of the electron's energy produces the crystal dilation, $\Delta(r)$ (fractional volume change). The electron's energy reads:

$$\varepsilon(k, r) = \varepsilon_0(k) + C \times \Delta r$$

Here C is constant with dimension of energy related to the deformation potential. The electron's energy is $\varepsilon_0(k) = \hbar^2k^2/2m_e$. The dimensionless dilation expresses through the phonon's amplitude, Δ_{ph}, and the phonon's wave vector q: $\Delta r = \Delta_{\text{ph}} \times q$. The phonon's amplitude (rather the measure of the lattice distortion) is $\Delta_{\text{ph}} = (\hbar/2M\omega_q)^{1/2}$ (as it is defined above). Here the frequency is, $\omega_q = c_s q$, where c_s is the speed of sound. The interaction Hamiltonian H (the deformation potential's perturbation) in the Born approximation for the scattering expresses through the matrix element of $C\Delta(r)$ for the transition between one-electron states k, and k' taking into account all states of electrons and phonons:

$$H \propto C\left(\frac{\hbar}{2M\omega_q}\right)^{1/2} \times q$$

The total coupling (collision) rate is calculated by the integration over the phonon phase space taking into account energy and quasi-momentum conservation (see Appendix J in Kittel, 1996,). The electron can transfer the energy only when the electron's group velocity exceeds the sound velocity, $v_g = \hbar k/m > c_s$. At the lattice temperature $k_BT_L >> \hbar c_s k$ (phonons occupation number is $n_q \approx k_BT_L/\ \hbar c_s k$) the energy transfer from electrons to phonons (to lattice) is proportional to the electron/atom mass ratio and to the electronic momentum, $\hbar k$

$$\nu_{\text{e–ph}}^{\text{energy}} \approx \frac{c^2 m_e k k_B T_L}{\pi \rho c_3^2 \hbar^3} = \frac{m_e}{M} \nu_{\text{e–ph}}^{\text{mom}} \propto \frac{m_e}{M} \varepsilon_e^{1/2} \tag{2.16}$$

Dependence on the electron's energy is the same as it follows from the kinetic considerations.

The electron–phonon energy transfer has been calculated at the electron temperature well in excess the Debye temperature of the lattice at $T_e >> T_L$, T_D (Kaganov et al., 1957). Rewriting the authors results in the modern notation the electron–phonon energy exchange rate takes a form similar to energy exchange rate in simple plasma (Landau, 1937):

$$\nu_{\text{e-ph}}^{\text{en}} \approx b \cdot \left(\frac{m_e}{M} \right) \nu_{\text{e-ph}}^{\text{mom}} \tag{2.17}$$

Here $\nu_{\text{e-ph}}^{\text{mom}} = v_e / l_{\text{mfp}}$, and b is a constant that should be found from comparison to the experiments.

Three decades later, Allen (Allen, 1987) calculated the electron–phonon energy transfer rate from the degenerated electron gas in a metal at $\varepsilon_F > T_e > T_L > T_D$ with the help of the kinetic equations employing Bloch–Boltzmann–Peierls collision integral.

We omit the electron mass renormalization factor at the temperature above the Debye limit on the basis of the experiments (Garl, 2008; Comins, 1972) and theory (Grimvall, 1981). Then Allen's electron–phonon energy transfer rate takes the following form:

$$\nu_{\text{e-ph}}^{\text{en}} \approx \pi \hbar \omega_{\text{ph}}^2 / \varepsilon_F \tag{2.18}$$

The explicit dependence on the electron temperature is absent except the small correction, decreasing with the growth of the electron's temperature. One can see that taking $\omega_{\text{ph}}^2 = m_e \omega_{\text{a}}^2 / M \varepsilon_F$, Eq. (2.18) is reduced to the form qualitatively similar to (2.16) but lacking the dependence on the electron's energy, $\nu_{\text{e-ph}}^{\text{en}} \approx \pi \frac{m_e}{M} \frac{h\omega_a^2}{\varepsilon_F}$

Therefore, we adopt for the electron–phonon energy transfer rate in the laser-excited good metals the formula (2.16) with the momentum transfer rate from (2.15) in the general form applicable for the good metals:

$$\nu_{\text{e-ph}}^{\text{en}} \approx \nu_{\text{e-ph}}^{\text{mom}}(T_{\text{room}}) \cdot \left(\frac{m_e}{M}\right) \cdot \left(\frac{k_B T_e}{eV}\right)^{1/2} \tag{2.19}$$

This formula for aluminium in an explicit form is presented below:

$$(\nu_{\text{e-ph}}^{\text{en}})_{\text{Al}} = 1.72 \cdot 10^{10} \left(\frac{k_B T_e}{0.025\text{eV}}\right)^{1/2} \quad (\text{s}^{-1})$$

At $k_b T_e \approx 10$ eV the above formula gives the energy transfer rate of $3.45 \times 10^{11}\ \text{s}^{-1}$ and equilibration time of 3 ps in qualitative agreement with experiments.

2.4.4 Electron-to-Ion Momentum and Energy Transfer

The intense short laser pulse transforms a metal to plasma in conditions when electrons acquire the energy close to the Fermi energy or ionization potential. We consider first the simple two-species (electrons and singly charged ions) plasma. The electron–electron, electron–ion, ion–ion interactions are governed by the Rutherford cross section for the scattering of charged particles on small angles (Lifshitz and Pitaevski, 1981). This approach is valid for the length scale much larger than the Debye screening distance, $l >> \lambda_D = v_e/\omega_{\text{pe}} \sim$ Angstrom, and for time intervals longer than inverse plasma frequency, $t >> \omega_{\text{pe}}^{-1} \sim 0.1$ fs. Obviously, these conditions are fulfilled during the short pulse laser–solid interaction where the pulse duration is in a range from several fs to a few hundred fs. The absorption depth is in the order of several hundred inter-atomic distances. The explicit formulae for all relevant collision rates are well known (Lifshitz and Pitaevski, 1981; Kruer, 1988). The important difference with the electron–phonon interactions is that the Coulomb cross section is inversely proportional to the squared electron energy. Therefore, the Coulomb collision rate decreases when the electron temperature grows up while in the electron–phonon collisions frequency grows up in proportion to the square root of the electrons' temperature. We present the electron–ion momentum transfer rate in a form useful for the practical estimates (Kruer, 1988):

$$\nu_{\mathrm{ei}} \approx 3 \times 10^{-6} \ln \Lambda_{\mathrm{ei}} \frac{n_e}{T_{\mathrm{eV}}^{3/2}} \tag{2.20}$$

Here the electron's energy, $k_B T_e$, is in electron-volts (eV), electrons' number density $n_e = Z \times n_a$; $\ln \Lambda_{\mathrm{ei}}$ is the Coulomb logarithm, which can be calculated as follows: $\Lambda_{\mathrm{ei}} = 1.53 \times 10^{10} \left[\left(\frac{(k_B T_e)}{n_e} \right] \right]^{1/2}$. The maximum collision rate at the low electron energy limit is set by the maximum collision rate allowed by the laws of Nature. Minimum mean free path for the electron–ion collision is the inter-atomic distance, d_a. Thus, the natural maximum for the electron–ion collision rate in solid plasma is $\nu_{\mathrm{ei,max}} \approx v_e / d_a$ (Allen, 2000; Eidmann et al., 2000). This limit for Al (n_a = 6 × 10^{22} cm^{-3}), d_a = $(3n_a/4\pi)^{-1/3}$ = 1.58 Å, and v_e = 2 × 10^8 cm/s equals 1.26 × 10^{16} s^{-1}. Correction on the non-ideal plasma state near the ablation threshold results in a decrease in the maximum rate (see below).

2.4.5 Electron-to-Ion Energy Exchange Time

The electrons and ions establish the Maxwell energy distributions within the each sub-system during the time much shorter than the time necessary for the energy transfer from electrons to ions. Therefore, the temperature equilibration proceeds between two groups of particles possessing different temperatures. The kinetic equations for electrons and ions are reduced to the equations for electron and ion temperature (two-temperature approximation) coupled via the electron–ion energy exchange term (sink in the electron energy and the energy source for the ions). Landau calculated this term analytically (Landau, 1937) treating the electron–ion collision integral in the approximation of a small change of momentum in a single collision that suffices the dominance of the far collisions in the Coulomb interactions. The calculation of the Landau collision integral with the Maxwell distributions for electron and ions produces the explicit form for the energy exchange rate in the coupled equations for electrons and ions temperatures. The equation for decrease in the electron energy density when the electron–ion energy transfer dominates and other losses are neglected reads

$$\frac{\partial Ee}{\partial t} \equiv \frac{3}{2} n_e \frac{\partial T_e}{\partial t} = -v_{\mathrm{ei}}^{(\mathrm{en})} \cdot n_e (T_{\mathrm{ei}} - T) \tag{2.21}$$

Accordingly, the energy sink in the above equation is the energy source for the ion energy density equation. The energy transfer rate from electrons to ions has the following form (Landau, 1937):

$$v_{\mathrm{ei}}^{(\mathrm{en})} \approx \frac{m_e}{M} v_{\mathrm{ei}}^{(\mathrm{mom})} \tag{2.22}$$

Note the close similarity to the electron–phonon energy transfer rate of Kaganov et al. (1957). The energy transfer time by the Coulomb collisions at conditions when the electron–ion momentum transfer rate is a maximum resides in a range of several picoseconds. For example, in Al, $m_e/M = 2 \times 10^{-5}$, $v_{\mathrm{ei}}^{(\mathrm{mom})} = 1.7 \times 10^{16}\ \mathrm{s}^{-1}$, the energy exchange time constitutes 3 ps.

The fact that electrons and ions energy distribution functions have the Maxwell distributions with the separate time-dependent temperatures and with the temperature-dependent energy exchange rate is the basis of two-temperature (two-fluid) description of plasma that has been understood long ago (Landau, 1937). Later on Kaganov et al. (1957) introduced the similar approach for calculation of the electron–phonon energy exchange and two-temperature approximation for description electron–phonon interaction in a metal.

2.4.5.1 Non-ideality effects

In the ideal plasma considered above the thermal energy of electrons exceeds the energy of Coulomb interaction. In order to check the applicability of the ideal plasma formulae for description of the laser-affected metal let us estimate the number of electrons in the Debye sphere (sphere with the radius equal to the

Debye screening length), $D_{\mathrm{De}} = 1.7 \times 10^9 \left(\frac{\varepsilon_e^3}{n_a \cdot Z}\right)^{1/2}$. Here the electron energy, ε_e, is in eV, the number density is in cm^{-3}. The criterion of the ideal plasma is $N_{\mathrm{De}} > 1$. In solid density Al plasma ($n_e = n_a \times Z = 1.86 \times 10^{23}$ cm^{-1}; $\varepsilon_e = 10$ eV) one gets $N_{\mathrm{De}} = 0.126$ that is indication on the dominance the Coulomb energy contribution over the thermal energy. Cauble and Rozmus (1995) established the relation of the electron–ion momentum exchange rate in the non-ideal two-component plasma (electrons and ions) to plasma frequency and coupling parameter, N_{De}, as follows:

$$\frac{\nu_{\mathrm{ei}}^{\mathrm{mom}}}{\omega_{\mathrm{pe}}} = \frac{\ln \Lambda_{\mathrm{ei}}}{10 \cdot N_D} \tag{2.23}$$

For Al at the above conditions taking $\Lambda_{\mathrm{ei}} = 9N_D$, one gets $\ln\Lambda_{\mathrm{ei}} \approx 0.126$ and $\frac{\nu_{\mathrm{ei}}^{\mathrm{mom}}}{\omega_{\mathrm{pe}}} = 0.1$ making the collision rate well below its natural limit in the ideal plasma case.

The significant difference in electron and ion temperature is typical for ultra-short laser–matter interaction when electrons are heated to the level of several electron-volts during the pulse time while the ions remain cold. The difference in the electron and ion energy results in the increased screening that in turn decreases the collision rate. The decrease in the collision rate in the strongly coupled plasma may affect two important processes: the inverse bremsstrahlung absorption process and the processes of establishing the equilibrium energy distribution that in turn also affects absorption.

2.4.5.1 Effects of the oscillations in high-frequency electromagnetic field on the electrons' collision rate

Conventionally the electrons' oscillations and collisions are considered to be independent assuming that the electron's kinetic energy significantly exceeds the oscillation energy of an electron in the field, $\varepsilon_{\mathrm{th}} >> \varepsilon_{\mathrm{osc}}$. V. P. Silin (Silin, 1965) found the general solution for the problem of the coupling of oscillations with collisions in the strong high-frequency electric field. His solutions, obtained with the help of the kinetic equation for plasma in a strong, uniform electric field, are valid from low to high oscillation energy ($\varepsilon_{\text{osc}} >> \varepsilon_{\text{th}}$) for the linear and circular polarization. We present an approximate expression for the collision rate including the oscillation energy, which gives the physically correct limits for low and high oscillation energy coinciding with the correct solution (Silin, 1965).

Let us consider the interaction of the short intense laser beam in condition when the kinetic and oscillation energy of electron are comparable. From the scaling the oscillation energy with intensity and wavelength, $\varepsilon_{\text{osc}} = 9.31 \cdot (I/10^{14}\ \text{W/cm}^2) \cdot \lambda^2_{\mu\text{m}}$ [eV], one can see that at intensity around 10^{13} W/cm^2, and with the laser wavelength around micron, the oscillation energy reaches electron-volt and might be comparable to the electrons' thermal energy.

The coupling of the electric charge kinematics to the oscillations in the high-frequency electric field can be seen explicitly in the Hamiltonian approach to the interaction problem (Silin, 1965). The canonical momentum of a charged particle is the sum of the kinematic momentum and the oscillation momentum depending on the vector potential of the field:

$$\bar{p} = m_e \bar{v}_e + \frac{e}{c}\bar{A}$$

The vector potential, A, is routinely connected to the electric field:

$$\bar{E} = -\frac{1}{c}\frac{\partial \bar{A}}{\partial t}; \quad \bar{A} = -\frac{c}{i\omega} E_o \cdot \exp(i\omega t)$$

The second term in the formula for the canonical momentum is the conventional form of the oscillation momentum following from the Newton equation for the free electron's oscillation in the HF field (see Chapter 1). Let us qualitatively analyse the electron's state in between collisions. Free electron moves with the transport velocity, v_e, in the absence of the HF field. When the field is switched on, the oscillations are overlapped with the transport motion. Electron starts oscillating in the direction, x, of the field at the moment of switching the laser on, t_o. One can reasonably assume that the energy of electron averaged over the many periods under assumption that, $t - t_o >> \omega^{-1}$, is the sum of the kinetic and oscillation energy, $\varepsilon_e = \varepsilon_{\text{th}} + \varepsilon_{\text{osc}}$.

We estimate the collision rate using the classical approach because the oscillation energy is much larger than the photons energy, $\varepsilon_{\text{osc}} > \hbar\omega$. At the collision's moment, the total electron's energy is $\varepsilon_e = \varepsilon_{\text{th}} + \varepsilon_{\text{osc}}$. Therefore, the total energy enters into the Coulomb cross section and to the effective collision rate as follows:

$$\nu_{\text{eff}} \approx \frac{\nu_{0,\text{ei}}}{\left(1 + \varepsilon_{\text{osc}} / \varepsilon_{\text{th}}\right)^{3/2}} \tag{2.24}$$

Here $\nu_{0,\text{ei}}$ is the conventional collision rate where the oscillation energy in the HF field is neglected. As a consequence, the electrons' collision rate in the strong field depends on the field intensity, frequency and polarization decreasing with the increase of the oscillation energy. Therefore, the account for the oscillation energy in the collision rate leads to the decrease in the energy absorption due to inverse bremsstrahlung, while the electronic heat conduction increases.

2.5 Modification of the Electron Distribution Function: From the Fermi–Dirac to the Maxwell–Boltzmann

A metal is heated to the high electron temperature $k_B T_e \leq \varepsilon_F$ remaining intact. The atomic bonds are broken and a metal is in state of solid-state density plasma. However, the material properties of laser-affected metal continue to be modified. The electron distribution function changes from the Fermi–Dirac distribution to the main part of the Maxwell–Boltzmann distribution.

The electron and lattice temperature equilibrates in a few picoseconds after the end of the pulse. To the moment of temperature equilibration the ionic distribution as well as the electron distribution is not in a complete equilibrium: the distribution lacks of the high-energy tail that needs longer time to be formed.

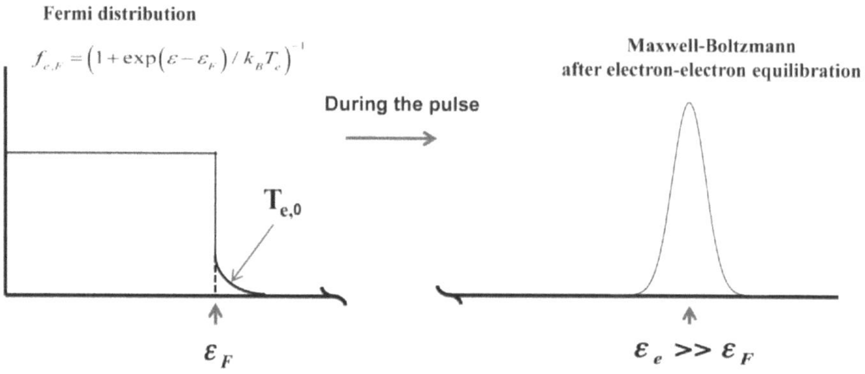

Figure 2.3 Transformation of the electrons distribution function during the ultra-fast laser excitation of a metal.

Building up the high-energy tail in the Maxwell distribution

It is known that atoms from the high-energy tail of the Maxwell distribution play crucial role in such equilibrium processes as melting (Fecht, 1992) and evaporation (Landau and Lifshitz, 1980). The proper concentration of thermal point defects (7–10%) is crucial for the initiation of the equilibrium melting at the melting point, T_{melt}. The energy for formation of such defect is equal to approximately 10 × T_{melt}. Therefore, in order to create such defects at the melting point the atoms with such energy should be generated. Similarly in order to evaporate an atom in equilibrium, i.e. to remove it from a solid, one needs delivering to that atom energy in excess of the binding energy, ε_b, of a few electron-volts. Conventionally the binding energy per atom is 10–20 times larger than the vaporization energy. Thus, melting at the melting point and vaporization at the vaporization temperature in equilibrium could proceed only if the full Maxwell distribution, including the high-energy tail in a range of 0.1–5 eV, is established.

In order to describe transient material state in a metal relaxing after the ultra-short-pulse excitation the knowledge of the time history for evolution and the establishing of the full Maxwell distribution is necessary. In the collision between particles of the same mass the whole particle energy can be transferred to other one in a single collision. During the energy transfer time for the electron–ion temperature equilibration the main part of the Maxwell distribution involving 80–90% of atoms, but lacking the high-energy tail has been established.

Particle needs more successive collisions in order to gain energy exceeding the average one, i.e. to build up the high-energy tail of the distribution. This process can be considered as diffusion along the energy axis (Uhlenbek and Ornstein, 1930; Lifshitz and Pitaevski, 1980). The diffusion coefficient, D, conventionally expresses through the energy gain rate, $\frac{d\varepsilon}{dt}$, as follows:

$$D_\varepsilon \approx \varepsilon \cdot \frac{d\varepsilon}{dt} [(\text{Joule})^2/\text{sec}]$$

Time to reach the energy, ε, during the diffusion along the energy axis reads $t_\varepsilon = \frac{\varepsilon^2}{D_\varepsilon}$. The energy gain rate for a particle, increasing its energy solely in collisions with the counterparts of the same nature, is proportional to the collision rate, $\frac{d\varepsilon}{dt} = \varepsilon \,.\, \nu$. Note that in collisions of the identical particles the momentum and energy transfer rates coincide. Therefore, the time to reach energy, ε, is inversely proportional to the collision rate at the energy ε.

$$t_\varepsilon = (\nu^{\text{mom}}(\varepsilon))^{-1}$$

It was found by MacDonald et al. (1957) that the time needed for establishing the *high-energy tail* of the electron distribution in plasma is much longer than that necessary for establishing a main part of distribution with temperature T_e. The momentum exchange rate for the Coulomb collision is $\nu_{\text{ei}} \propto T_e^{-3/2}$. The time for electron to reach the energy $\varepsilon_e >> k_B \times T_e$ is sufficiently longer (see MacDonald et al., 1957):

$$t_{\text{tail}}^{\text{el}} / t_{\text{main}}^{\text{ei}} = (\varepsilon_{el}/k_B T)^{3/2} \tag{2.25}$$

Similarly the time for establishing the full Maxwell distribution in the ion-subsystem after the temperature equilibration reads, $t_{\text{tail}}^{\text{ion}} / t_{\text{main}}^{\text{ei}} = (\varepsilon_{\text{ion}} / k_B T)^{3/2}$. One can see that the time for establishing the high-energy tail with $10 \times k_B T$ is $10^{3/2} = 31.6$

times longer than the temperature equilibration rate and constitutes around 100 ps for aluminium.

It appears that the formation of the full Maxwell distribution in the surface layer differs from the process in the bulk. The atoms in the outermost surface layer next to the vacuum are in fact in a different condition compared to the atoms in the bulk. It is well- known that the surface atoms are loosely bound to the bulk with the part of bonds dangling or saturated with foreign atoms (Zangwill, 1988; Prutton, 1994). The effects of different bonding at the surface leads to decreases in the Debye and melting temperatures; to changes in the bond length and inter-atomic distance as well as the crystalline structure and nature and rate of any phase transition. Prutton noticed in 1994 that "*... many surface phases are actually metastable, i.e. the surface is not in a true thermodynamic equilibrium*". The energy distribution in the outermost surface layer is the important characteristic affecting the removal of atoms from the surface layer at the ablation threshold. The energy distribution is responsible for the relative contribution of non-equilibrium ablation and thermal evaporation. We will consider this problem in relation to the ablation thresholds in ambient gas and in vacuum in Chapter 5.

2.6 Electronic Heat Conduction

Heat transport mechanism continuously changes during the swift excitation by the short intense laser pulse and after the end of the pulse. Let us follow these changes from low temperature until the conversion to plasma.

Conventionally, the heat flux has a diffusion form when a mean free path of a heat carrier is much less than the space scale of an area under consideration:

$$\bar{q}_{\alpha} = k \cdot \nabla T \tag{2.26}$$

The heat conduction coefficient expresses through the specific heat, C, number density, n, and diffusion coefficient, D, for the carriers:

$$\kappa = C_e \cdot n_e \cdot D\left[\frac{\mathrm{W}}{\mathrm{K}\cdot\mathrm{cm}}\right] \tag{2.27}$$

The diffusion coefficient from the kinetic theory is the function of the carrier velocity and the mean free path, while the mean free path expresses through momentum transfer collision rate, $l_{\text{mfp}} = v_e/v^{\text{mom}}$, thus giving

$$D = \frac{l_{\text{mfp}} v}{3} = \frac{v^2}{3v^{\text{mom}}}[\text{cm}^2/\text{sec}]. \tag{2.28}$$

The characteristic time for the cooling of the absorption layer of the size of l_{abs} follows directly from the heat conduction equation:

$$t_{\text{cool}} \approx \frac{l_{\text{abs}}^2}{D} \tag{2.29}$$

Let us consider the heat transport at the different levels of excitation at relatively low temperature, in non-ideal plasma and in the ideal plasma separately.

1. Low excitation $\varepsilon_F >> k_B T_e > k_B T_L > k_B T_D$

   The electron–phonon momentum exchange rate may be adjusted to the experimentally measured optical data at the room temperature, $v_{\text{e-ph}}^{\text{mom}} \propto T_e^{1/2}$. The heat diffusion coefficient according to (2.28) then also grows up as $D \propto T_e^{1/2}$. However, the heat conduction coefficient at low temperature, $k_B T_e < 3\varepsilon_F/\pi^2$, when the electrons are in the degenerate state with the heat capacity, $C_e = \pi^2 k_B^2 T_e / 2\varepsilon_F$, is the strong function of the electron temperature:

$$\kappa_e \propto T_e^{3/2} \text{ at } k_B T_e < 3\varepsilon_F/\pi^2 \tag{2.30}$$

In plasma state the heat conduction coefficient is

$$\kappa_e \propto T_e^{1/2} \text{ at } k_B T_e < 3\varepsilon_F/\pi^2 \tag{2.31}$$

2. *Electronic heat conduction in plasma*

   At the temperature equivalent to several eV electrons could be treated as those in the ideal plasma. In the ideal electron–ion plasma the heat capacity is as for the ideal gas, $C_e = 3k_B/2$ and electron–ion momentum exchange rate $\nu_{ei} \sim (T_e)^{-3/2}$. Thermal diffusion coefficient is the strong function of the electron temperature:

$$D = v_e^2/\nu_{ei} \propto T_e^{5/2} \tag{2.32}$$

   For example, in aluminium at $k_B T_e = \varepsilon_F$, the diffusion coefficient is $D \approx 0.9$ cm^2 s^{-1} approximately twice larger than at room temperature. The absorption length in Al is of the order of $l_{abs} \approx 40$ nm. Therefore, the cooling time is $t_{cool} \approx (l_s)^2/D \approx 16$ ps affecting other processes in the laser-affected metal.

3. *Effects of the polarization on the electronic heat conduction in HF electric field (Gamaly and Rode, 2016)*

   Let us consider the effect of the monochromatic polarized laser beam on the electronic heat flow in plasma. Conventional heat flow is a vector directed along the temperature gradient coinciding with the vector of transport velocity of electrons. The heat conduction flow in the electric field should include two true vectors, the field vector E, and the velocity vector v (Lifshitz and Pitaevsky, 1981). The heat conduction coefficient in this case should be a tensor and the heat flow has a form

$$q_\alpha = -\kappa_{\alpha\beta} \frac{\partial T}{\partial x_\beta}.$$

The general expression of second-order tensor constructed from the true vector, E, is

$$\kappa_{\alpha\beta} = \kappa_1 \delta_{\alpha\beta} - e_\alpha e_\beta \kappa_2; \; \bar{e} = \frac{\bar{E}}{E}. \tag{2.33}$$

Here e is the unit vector in the direction of the electric field; κ_1 and κ_2 are two scalar functions, which can be obtained from the solution of the kinetic equation. The heat flow in the presence of the field reads

$$q_\alpha = -\kappa_{\alpha\beta}\frac{\partial T}{\partial x_\beta} = \kappa_1\frac{\partial T}{\partial x_\alpha} - \kappa_2 e_\alpha e_\beta \frac{\partial T}{\partial x_\beta}. \tag{2.34}$$

Note that field-related second term equals zero when the polarization direction is perpendicular to the temperature gradient. Let us now consider the effects of linear and circular polarizations in laser--plasma interactions.

The intense laser pulse hitting the solid surface along the normal (z coordinate) produces hot plasma in the circular spot (x,y plane) of radius r_{foc}, surrounded by cold unperturbed target. The temperature has maximum in the centre of the circle while the temperature gradient is a maximum at the outer boundary of heated spot directed along the radius from the centre outside.

Linear polarization. Let the laser electric field direction be along x-axis. Then the field-affected heat flow part reads

$$q_{\text{field}} = \kappa_2 e_x^2 \frac{\partial T}{\partial x}$$

Note that field-affected flow exists in both directions only along the x-axis, while in the other parts of heated circle heat flow is unaffected. Therefore, the expected effect of the linear polarization is the elongation of the laser-affected spot in the polarization's direction, turning a circle into an oval with the large axis aligned along the direction of the electric field.

Circular polarization. The electric field vector of the incident beam at any moment of time is collinear to the temperature gradient directed along the radius of the beam. Thus, the field influence on a heat flow is evenly distributed along the circle embracing the heated zone and leading to the change of the zone's radius.

The coupling between tight focusing, polarization and heat conduction was observed in the experiments of the direct laser writing (Malinauskas et al., 2015). It was demonstrated that this coupling affects the size, shape and absorption in the laser-affected area on the surface of different materials. Here is also effect of anisotropic absorption in the case of tight focusing: p- and s-polarized parts of laser beam are absorbed differently (Tikhonchuk, 2020, private communication).

2.7 Summary

A metal stays intact, and the number density of conduction electrons remains unchanged during the short intense pulse action. However, the optical and material properties of the laser-affected metal are continuously changing during the interaction along with the increasing intensity and growing up electron temperature.

- The laser wave penetrates inside a metal to the length (e-fold exponential decay of the electric field amplitude) $l_s = \frac{c}{\omega \cdot \kappa}$; $\sqrt{\varepsilon} = n + i\kappa$
- Electronic heat capacity changes from that for the degenerate state to the ideal gas value. Lattice heat capacity changes from that for the solid ($3k_B$ per atom) to that for the ideal gas in plasma state ($3k_B/2$ per atom).
- The electron–phonon momentum and energy exchange rates at $k_B T_e \leq \varepsilon_F$ are as follows:

$$\nu_{\text{e-ph}}^{\text{mom}} = \nu_{\text{e-ph}}^{\text{mom}}(T_{\text{room}}) \cdot \left(\frac{k_B T_e}{0.025\ \text{eV}}\right)^{1/2} \quad \nu_{\text{e-ph}}^{\text{mom}} \approx \frac{m_e}{M} \nu_{\text{e-ph}}^{\text{mom}}(T_e)$$

  Electron-to-ion momentum and energy transfer at $k_B T_e > \varepsilon_F$ are as in the ideal plasma
- The electron distribution function at $k_B T_e > \varepsilon_F$ changes from the Fermi–Dirac to the Maxwell distribution.
- The longest relaxation time is the time for establishing high-energy tail in the Maxwell distribution $\frac{t_{\text{tail}}^{\text{ion}}}{t_{\text{main}}^{\text{ei}}} = \left(\frac{\varepsilon_{\text{ion}}}{k_B T}\right)^{3/2}$, $\varepsilon_{\text{ion}} >> k_B T$.

Chapter 3

Interaction with Dielectrics

The optical properties of dielectrics in a high-frequency field experience dramatic changes when the imposed electric field's amplitude gradually increases. At low intensity, the imaginary part of dielectric permittivity is close to zero (no absorption) and solid is transparent. At elevated intensity, the threshold-free multi-photon ionization process starts transferring electrons from the valence band to the conduction band. The incident light starts interacting with the conductivity electrons via inverse bremsstrahlung process increasing the electrons energy. On the language of classical physics electron scattered in the field of ion absorbs photon. Three-body participation in the process is necessary for the momentum conservation. A free electron gaining the energy above the ionization barrier becomes capable to knock out a bound electron into the conduction band (ionization by the electron's impact).

(1) Essential difference of the light wave interaction with the initially transparent dielectrics from that with metals is the following. The intense short beam interacting with a transparent dielectric creates the electrically inhomogeneous medium. That means that the laser-modified permittivity varies in space mapping the spatial

Femtosecond Laser-Matter Interactions: Solid-Plasma-Solid Transformations at the Extreme Energy Density
Eugene Gamaly

ISBN 978-981-4877-40-4 (Hardcover), 978-1-003-25661-8 (eBook)
www.jennystanford.com

intensity distribution of the pump pulse. The spatial distribution of the material density remains intact and homogeneous because the pulse duration is shorter than the energy transfer time from electrons to the lattice. At certain threshold of absorbed energy, the number density of electrons in the conduction band rises to the level when the real part of the dielectric function turns to zero.

(2) At this stage of interaction, the resonant excitation of plasma waves near zero point (resonance absorption) is possible when the component of electric field coincides with the permittivity gradient. However, the difference with the similar phenomenon in plasma relates to the fact that almost 90% of valence electrons still occupy the valence band creating the electrostatic field affecting the interaction.

(3) The electrons collision rate grows up along with the increase of the electrons number density approaching to the rate's maximum. The Coulomb collisions dominate in a dielectric converted into plasma. The solid is converted to the metal-like state, remaining still intact, and the laser wave becomes evanescent and absorbing. During the following heating, the electrons can be promoted to the continuum transforming dielectric to plasma state. Along with the transformation process from solid to plasma, the transient optical and material properties experience many changes. In what follows, we consider step by step the processes of dielectric-to-plasma transformation when a solid still remains intact. After the energy transfer from electrons to ions, the hydrodynamic motion becomes possible (see Chapter 5).

3.1 Ionization in the Strong High-Frequency Electric Field: Electrons Transfer from the Valence to Conduction Band and to Continuum

Bound electron should overcome the potential barrier equal to the band gap energy, Δ_g, in order to be transferred from the valence band to the conduction band. The width of this barrier depends on the force of applied electric field as follows:

$$\Delta l = \Delta_g / eE \qquad (3.1)$$

The electron in high-frequency (HF) field oscillates with the velocity, $v_{osc} = eE/m_e\omega$. The time to cross the barrier reads, $t_{bar} = \Delta l / v_{osc} = \Delta_g / 4\varepsilon_{osc} \cdot \omega$. Here, the energy of the electron oscillation in the linear polarized field was introduced, $\varepsilon_{osc} = e^2E^2/4m_e \cdot \omega^2$ (see Chapter 1). Comparing this time with the oscillation period of the HF field, $T = 2\pi/\omega$, one can see that two interaction regimes of the HF field interaction with electron confined in the potential well of depth, Δ_g, can be introduced. In the tunnelling regime when $T >> t_{bar}$ (or $\omega_t = (2\pi/t_{bar}) >> \omega$) the ionization depends on the instantaneous field (intensity) value and does not depend on the frequency. In the multi-photon regime (MPI), when $T << t_{bar}$, the electrons have no time to pass through the barrier during one cycle (Keldysh, 1965). Correspondingly, the value of the parameter combining the laser and material properties, the Keldysh parameter, $\gamma^2 = \Delta_g / 4\varepsilon_{osc} = 2\pi t_{bar} / T$, defines the range for dominance of the different ionization mechanisms. The MPI process dominates at $\gamma > 1$, while the tunnel ionization occurs at $\gamma < 1$ ($\varepsilon_{osc} > \Delta_g$).

Keldysh calculated the transition between the bound states in the low order of the perturbation theory. Therefore, the matrix element is proportional to $\Delta_g/\hbar\omega$. Thus, in the transfer of electrons from the valence band to the conduction band calculated in a similar way the energy of photon is also assumed to be much lower than the band gap, $\hbar\omega/\Delta_g << 1$. The ionization rate calculated under these assumptions is inapplicable for ionization of the narrow band gap dielectrics, such as silicon, by UV light when $\hbar\omega \geq \Delta_g$. In the case, when one of the high harmonics of the incident monochromatic wave is close to resonance with the electronic transition to s-state, then the two-step process is possible.

Let us define the laser parameters for interaction with the wide-band gap dielectrics such as sapphire or silica ($\Delta_{g,Sp} = 9.8$ eV; $\Delta_{g,SiO2} = 7.8$ eV) in both regimes using for the oscillation energy for the linearly polarized light in the form (see Chapter 1)

$$\gamma^2 = \Delta_g/37.24 \cdot (I/10^{14}\ \mathrm{W/cm^2}) \cdot \lambda^2_{\mu\mathrm{m}}. \tag{3.2}$$

For sapphire at λ = 0.8 μm, one gets γ^2 = 0.4 at $I = 10^{14}$ W/cm^2 and for silica γ^2 = 0.327. Hence, the tunnelling ionization regime dominates in ionization of the wide-band gap dielectrics at intensity ≥ 10^{14} W/cm^2.

3.1.1 Tunnelling Ionization Rate in the Limit $\gamma << 1$

3.1.1.1 Linear polarization

The probability of the tunnelling ionization per unit time, at $\gamma << 1$ (static limit), reads (Landau and Lifshitz, 1980)

$$w_{\mathrm{tun}} \approx 3.91 \cdot \omega_{\mathrm{at}} \cdot \left(\frac{I}{I_{\mathrm{at}}}\right)^{1/4} \cdot \exp\left\{-\frac{2}{3}\left(\frac{I_{\mathrm{at}}}{I}\right)^{1/2}\right\} \tag{3.3}$$

Here, I_{at} = 3.4 × 10^{16} W/cm^2 is the atomic intensity, and $\omega_{\mathrm{at}} = J_{\mathrm{at}}/\hbar$ = 2.06 × 10^{16} s^{-1} is the atomic frequency. It was indicated (Ammosov et al., 1986) that actually the quasi-static limit is achieved at $\gamma \le 1/2$. Indeed the time for the passing the barrier reads $t_{\mathrm{bar}} = \gamma^2/\omega$. Thus, at γ = ½ time for the tunnelling through the barrier equals $t_{\mathrm{bar}} \approx T/8\pi$ justifying that the electron experiencing practically the maximum amplitude of the field (see Appendix 8). It is instructive estimating the tunnelling rate at $I = 10^{14}$ W/cm^2 for the further comparison with the rates of different ionization mechanisms. One gets from (3.3) w_{tun} (10^{14} W/cm^2) = 3.16 × 10^{11} s^{-1}. Keldysh derived the formula (formula (40) in Keldysh, 1965) for the tunnelling probability for ionization of the dielectric with the band gap Δ_g at $\gamma << 1$ keeping small frequency dependent term in the exponent. We rewrite the Keldysh formula making it easy for comparing to (3.3) introducing $\varepsilon_{\mathrm{at}} = e^2/a_B$; $E_{\mathrm{at}} = e/a_B^2$ and γ, $\varepsilon_{\mathrm{osc}}$ as above:

$$w_{\mathrm{tun}} \approx \frac{\Delta_g}{\hbar} \cdot \left(\frac{\varepsilon_{\mathrm{at}}}{\Delta_g}\right)^{3/4} \left(\frac{I}{I_{\mathrm{at}}}\right)^{1/4} \cdot \exp\left\{-\frac{\pi}{2}\left(\frac{\varepsilon_{\mathrm{at}}}{\Delta_g}\right)^{3/2} \left(\frac{I_{\mathrm{at}}}{I}\right)^{1/2} \cdot \left(1-\frac{\gamma^2}{4}\right)\right\} \tag{3.4}$$

We omitted the term $(m_e \cdot \Delta_g/\hbar^2)^{3/2}$ [cm^{-3}] in the Keldysh formula reducing it to the probability per unit time. (Space dependence originates from the integration over the quasi-momentum space, $p^2 = m_e \times \Delta_g$.) Only correction on the difference between the band gap and atomic energy scale is introduced.

3.1.1.2 Tunnel ionization in the elliptically polarized electric field (Perelomov et al., 1966; Ammosov et al., 1986)

Elliptically polarized field has a form, $E = E_0\ (e_x \times \cos \omega t \pm \alpha \times e_y \times \sin \omega t)$. Here, $-1 \leq \alpha \leq 1$ stands for the linear polarization, and $\alpha = \pm 1$ stands for the circular polarization case. It was found that the probability of ionization in the elliptically polarized HF field is connected to the probability of ionization in the constant field (Perelomov et al., 1966). Moreover, it appears that the probability of ionization in the purely circular polarized field equals that in the constant field, $w_{\text{circ}} = w_{\text{const}}$. The probability of the tunnel ionization in the circular polarized field in the quasi-classical limit ($\hbar\omega << \Delta_g$, $E << E_{\text{at}}$) appears to be much larger than for the linear polarization, $(w_{\text{circ}}/w_{\text{lin}}) >> 1$. On the other hand in the tunnel regime the condition holds, $\varepsilon_{\text{osc}} >> \Delta_g$. The oscillation energy of electron in the HF field reads, $\varepsilon_{\text{osc}} \sim (1 + \alpha^2)$. Therefore, $\varepsilon_{\text{osc}}^{\text{circ}} = 2\varepsilon_{\text{osc}}^{\text{lin}}$ in the electric field of the same amplitude making the above result qualitatively understandable. As we see later, the probability of multi-photon ionization in the circular polarized light appears to be less than that for the linear polarization.

3.1.2 Multi-Photon Ionization Rate

3.1.2.1 Linear polarization

Keldysh calculated the probability of electron transition from the valence band to the conduction band in dielectric under the action of the linear polarized laser beam in the multi-photon limit at $\gamma >> 1$ ($\Delta_g >> \varepsilon_{\text{osc}}$) (Keldysh, 1965). The probability per unit time and unit volume reads in the simplified form using the original Keldysh's notations as follows:

$$w_{\text{MPI}} = \frac{2}{9\pi}\omega\left(\frac{m_e\omega}{\hbar}\right)^{3/2} \cdot \left(\frac{\varepsilon_{\text{osc}}}{4\Delta_g}\right)^{n_{\text{ph}}} \cdot \Phi\left(\left[2n_{\text{ph}} - \frac{2\bar{\Delta}}{\hbar\omega}\right]\right) \cdot \exp\{2n_{\text{ph}}(1-\varepsilon_{\text{osc}}/\Delta_g)\} \tag{3.5}$$

Here, the average number of absorbed photons, $n_{\text{ph}} = \left\langle \frac{\bar{\Delta}}{\hbar\omega} + 1 \right\rangle$, and the effective band gap, $\bar{\Delta} = \Delta_g + \varepsilon_{\text{osc}}$, are introduced. The angular bracket denotes the integer part of the number in brackets. Note that the space dependence in Eq. (3.5), $\left(\frac{m_e\omega}{\hbar}\right)^{3/2}$ follows from the integration over the quasi-momentum, $p^2 = \hbar\times\omega\times m_e$. Raizer (Raizer, 1976) presented the MPI rate for the ionization of a single atom per unit time in the scaling form

$$w_{\text{MPI}} = B \cdot \omega \cdot n_{\text{ph}}^{3/2}\left(\frac{\pi e^2}{m_e c}\frac{I}{\omega^2 \cdot J_0}\right)^{n_{\text{ph}}},$$

where J_0 is the ionization potential. The MPI probability for the ionization of dielectric per unit time from Eq. (3.5) can be also presented as a function of laser (intensity and frequency) and material (band gap) parameters:

$$w_{\text{MPI}} = B \cdot \omega\left(\frac{9.375 \cdot \lambda_{\mu\text{m}}^2}{\Delta_{g(\text{eV})}}\right)^{n_{\text{ph}}} \cdot \left(\frac{I}{10^{14}\ \text{W/cm}^2}\right)^{n_{\text{ph}}} \tag{3.6}$$

Here, the band gap is eV and laser wavelength is in microns. Let us estimate the MPI rate in the fused silica ($\Delta_g = 7.6$ eV) by the action of laser with $\lambda = 800$ nm light ($\hbar\omega = 1.55$ eV; $\omega = 2.356 \times 10^{15}$ s^{-1}). Thus, one gets $n_{\text{ph}} \approx 7.6/1.55 + 1 = 5.9$.

Note that the applicability of the used approximation ($\Delta_g >> \varepsilon_{\text{osc}}$) becomes questionable at $I \geq 10^{14}$ W/cm^2 ($\varepsilon_{\text{osc}} = 6.24$ eV).

The following scaling has been used for the calculation of the ionization probability of fused silica with good agreement to experimental data (Stuart et al., 1995, 1996):

$$w_{\text{mpi,SiO}_2} = 1.3 \times 10^{10} \cdot \left(\frac{I}{10^{13}\ \text{W/cm}^2}\right)^5 \tag{3.6a}$$

Equation (3.6) with n_{ph} = 5; λ = 800 nm; Δ_g = 7.6 eV, takes the form $w_{\text{mpi,SiO}_2} = 0.87 \times 10^{10} \cdot \left(\frac{I}{10^{13}\ \text{W/cm}^2}\right)^5$ close to that from the reference (Stuart et al., 1995, 1996), supporting the viability of the scaling by Eq. (3.6).

3.1.2.2 Time dependence of the ionization degree produced by the Gaussian laser pulse with the MPI domination

The MPI is the threshold-free mechanism exclusively generating electrons with the growing intensity until the electron impact ionization begins contributing. The rate equation for the number density of electrons produced by the only MPI is as follows:

$$\frac{dn_e}{dt} = (n_a - n_e) \cdot w_{\text{MPI}}$$

One can integrate the electron rate equation analytically taking MPI rate in the power form of Eq. (3.6), $w_{\text{MPI}} = \beta_m \cdot \left[\frac{I_0}{I} f(t/t_p)\right]^m$ $[s^{-1}]$.

Time-dependence of intensity is in the Gauss form, $f(t) = \exp\{-4(1-2t/t_p)^2\}$. The direct integration produces the time dependence of the ionization degree (n_e/n_a) in an explicit form as a function of laser and material parameters (see Appendix 9). The easily accessible tables for the error function make the practical estimates quick and handy. The small ionization degree to the end of the pulse, $n_e(t_p)/n_a < 1$, has the following form:

$$\frac{n_e(t_p)}{n_a} = \frac{\sqrt{\pi}}{4m^{1/2}} \cdot t_p \cdot \beta_m \cdot (I_0/I_1)^m \qquad (3.6b)$$

The MPI process dominates when a dielectric transforms to absorbing medium with the ionization degree of a few per cents. The above formula gives the reasonable estimates in agreement with the experiments (see Appendix 9).

3.1.2.3 Circular polarization

Important effect directly related to the quantum properties of photon is strong dependence of multi-photon ionization rate on polarization (Gavrila, 1992 and references therein). It was noted in earlier studies that the ionization rates for circular polarization are several orders of magnitude smaller than those for linear polarization. The explanation of this effect directly relates to the angular momentum. The selection rule for the angular momentum dictates that for each absorbed photon the magnetic quantum number has to increase by one unit (measured in the Planck constant). The ground state usually has low angular momentum. This implies that after absorption of many photons the angular momentum may become very large. Thus, the final state separates from the origin by a strong centrifugal barrier. That causes very poor overlap of the wave functions of electron in the final and ground state, which results in the small ionization cross section.

The experimental results in earlier studies were, however, more controversial (see discussion in Temnov et al., 2006). First clear evidence for dominance of circular over linear polarization was demonstrated in the experiments on multi-photon ionization of Na in the 8.5 GHz microwave field (Gallagher, 1992). The experiments on ionization of sapphire and fused silica by 50 fs, 800 nm powerful laser clearly demonstrated the difference between circular and linear polarization in qualitative agreement with the theoretical predictions for the MPI (Temnov et al., 2006).

In both materials, the six-photon MPI rate is found to be significantly higher for the linear polarization of the pump pulses (Temnov et al., 2006). The ratio of cross sections for circular polarization to that of the linear polarization in fused silica is $\sigma_{\rm Si}^{\rm circ}/\sigma_{\rm Si}^{\rm lin} = 0.27$, while in sapphire it is $\sigma_{\rm sap}^{\rm circ}/\sigma_{\rm sap}^{\rm lin} = 0.15$, qualitatively supporting the observations.

3.1.3 Ionization by the Electron Impact

The electrons transferred to the conduction band by the HF field ionization process start interacting with HF field as the conductivity electrons in a metal do. Free electron oscillating in the high-frequency laser field gains the energy from the field via three body collisions involving laser photon, electron and atom (ion) in a solid in order to comply with the conservation of momentum. When the conduction electron gains energy above the band gap, then collision of this electron with the valence electron can promote it to the conduction band. This ionization process is called the ionization by the electron impact. As an electron gains energy more than twice of the band gap the next collision may produce another electron possessing energy in excess of the band gap and therefore capable of ionizing another valence electron. Thus, the impact ionization at high intensity transforms to the avalanche process. Let us first consider the continuous energy gain by electrons through collisions in the HF field in the quasi-classical approximation. The electrons energy growth rate reads (see Chapter 1)

$$\frac{dE_e}{dt} = \frac{\omega}{c}\varepsilon_{\mathrm{im}} \cdot I[\mathrm{W/cm^3}]. \tag{3.7}$$

Inserting the imaginary part of the dielectric permittivity in the general form (see Chapter 1) into (3.7),

$$\frac{dE_e}{dt} = \frac{\omega_{\mathrm{pe}}^2}{c \cdot (\omega^2 + \nu_{\mathrm{eff}}^{\mathrm{e}})}\nu_{\mathrm{eff}} \cdot I.$$

The number density of electrons, transferred through the band gap, Δ_g, into the conduction band per unit time—the impact ionization rate, W_{imp}—reads

$$W_{\mathrm{imp}} = \frac{1}{\Delta_g}\frac{dE_e}{dt} = \frac{dn_e}{dt} = \frac{1}{\Delta_g}\frac{\omega_{\mathrm{pe}}^2}{c \cdot (\omega^2 + \nu_{\mathrm{eff}}^{\mathrm{e}})}\nu_{\mathrm{eff}} \cdot I. \tag{3.8}$$

Rearrangement of the above formula reveals the physical nature of the phenomenon:

$$W_{\mathrm{imp}} = \frac{2\varepsilon_{\mathrm{osc}}}{\Delta_g}\frac{n_e \cdot \nu_{\mathrm{eff}}}{(1 + \nu_{\mathrm{eff}}^2/\omega^2)}[\mathrm{cm^2 s^{-1}}] \tag{3.9}$$

The energy of the electrons' oscillations in the HF field is transformed by the effective collisions into electrons' kinetic energy. When electron gains the energy in excess of the band gap, it is transferred into conduction band. The impact ionization becomes very effective when $\varepsilon_{osc} > \Delta_g$. Therefore, the impact ionization shall strongly compete with the MPI process at some high intensity level because the criteria of efficiency of both processes are opposite. From (3.9) follows the exponential nature of the process: the number density of electrons grows exponentially if the ionization rate per one electron is constant. However, the ionization rate per one electron strongly depends on the effective collision rate and intensity. Moreover, the collision rate itself is a function of the growing electrons number density, as we will see later. As a rule, the ionization rate per electron has been used as follows:

$$W_{\rm imp} = W_{\rm imp}/n_e = \eta \times I \,[{\rm s}^1] \tag{3.10}$$

Impact ionization rate's coefficient reads

$$\eta = \frac{1}{c \times \Delta_g \times n_{\rm cr}} \times \frac{\nu_{\rm eff}}{(1 + \nu_{\rm eff}^2/\omega^2)}. \tag{3.11}$$

Here, the critical electron density corresponding to the frequency of ω of the laser, $n_{\rm cr} = \frac{m_e \times \omega^2}{4\pi e^2}$, was introduced. The coefficient (3.11) is strong function of the collision rate, reaching the maximum at $\nu_{\rm eff} = \omega$. The collision rate changes by two orders of magnitude under the action of the short intense laser pulse transforming a solid from the initial state to plasma. In solid dielectrics, the collision rate at room temperature is close to the electron–phonon collision rate, $\nu_{\rm e\text{-}ph}\,(T_{\rm room}) \approx 3k_BT_{\rm room}/2\hbar = 6 \times 10^{13}\ {\rm s}^{-1}$, while the maximum value in the plasma state is close to that in metals, $\nu_{\max} \approx 5 \times 10^{15}\ {\rm s}^{-1}$. Let us estimate the coefficient for silica under the action 800 nm ($\omega = 2.356 \times 10^{15}\ {\rm s}^{-1}$) laser taking $\Delta_g = 7.8$ eV; $n_{\rm cr} = 1.745 \times 10^{21}\ {\rm cm}^{-3}$. The minimum is $\eta_{\min} = 0.92\ {\rm cm}^2$/Joule. The maximum, achieved at $\nu = \omega$, equals $\eta_{\max} = 17.68\ {\rm cm}^2$/Joule. In the calculations for silica, this coefficient was set constant and equal to 10 cm^2/J (Stuart et al., 1995, 1996). Note that the electrons impact ionization rate starts prevailing the MPI rate around the intensity 10^{13} W/cm^2.

3.1.3.1 Transition to the avalanche regime (Raizer, 1977; Yablonovitch and Bloembergen, 1972)

Let us now obtain conditions when the impact ionization transforms to the avalanche mode. Free electron heated during the time, τ_Δ, getting the energy in excess of the band gap and being able to knock out another valence electron to the conduction band, $\varepsilon_e \geq \Delta_g$. These two electrons after the time, τ_Δ, are able each producing two more electrons. Each of two newly produced electrons repeats the process, which can be described as $n_e \approx n_{\text{seed}} \times 2^{t/\tau\Delta}$. Probability of the knockout ionization estimates as the collision rate, $w_{\text{imp}} = n_{\text{at}} \times \sigma_{\text{at}} \times v_e \approx 6.6 \times 10^{15}\ \text{s}^{-1}$ for silica (taking $\sigma_{\text{at}} = \pi \times r_{\text{at}}^2$ and $v_e \approx 2 \times 10^8$ cm/s). Time for the heating electron to the energy $\varepsilon_e = \Delta_g$ can be estimated with the help of (3.7–3.9) as follows:

$$\frac{dn_e}{dt} \approx \frac{n_e}{\tau_\Delta} \tag{3.12}$$

The electron number density growth can be presented in the form

$$\ln n_e(t) \approx \int_0^t \frac{dt}{\tau_\Delta}.$$

Keldysh introduced the formula for the rate when the impact ionization in the avalanche mode dominates:

$$w_{\text{aval}} \approx (1.5 \times 10^{15}\,\text{s}^{-1}) \times \left({\varepsilon_e(t)}/{\Delta_g} - 1 \right)^2 \tag{3.13}$$

Note that in this formula, the electron's energy depends on the fluence achieved during the pulse to that moment. Electron energy can be estimated from the absorbed energy density of (1.25) (Chapter 1) as follows: $\varepsilon_e(t) = k_B \times T_e = \frac{4A \times F(t)}{3n_e \times l_s}$. Taking the numbers for singly ionized silica at the fluence in excess of ionization threshold but below ablation threshold, $A \approx 0.5$; $F \approx 1$ Joule/cm^2; $n_e \approx 6 \times 10^{22}$ cm^{-3}; $l_s \approx 50$ nm, one obtains $\varepsilon_e \approx 13.9$ eV. In these conditions, the impact ionization transforms into the avalanche mode.

The MPI is the threshold-free process. Therefore, MPI produces the seed electrons even at low intensity. The seed electrons are then multiplied by the impact ionization, making both processes strongly interrelated. The impact ionization rate overcomes the MPI rate in interaction of 800 nm light with silica at the fluence around 0.1 J/cm^2 (average intensity $\geq 10^{13}$ W/cm^2 during 100 fs pulse).

3.1.4 Distribution Function of Electrons Transferred to the Conduction Band in Dielectrics

The Maxwell–Boltzmann distribution is applicable for the description of the energy distribution of any particles if the total number of particles is much less than the number of available energy states. The criterion for the applicability of the Boltzmann distribution for the electrons with the number density, n_e, and the average energy, $k_B T_e$, is as follows (Landau and Lifshitz, 1980):

$$n_e \times \left(\frac{\hbar^2}{m_e \times k_B \times T_e} \right)^{3/2} << 1 \tag{3.14}$$

The term in the brackets in the above formula is the volume in the phase space. One can see that electrons transferred into the conductivity band by swift ionization of dielectric comply with the above criterion well up to full conversion to plasma ($n_e \sim 10^{23}$ cm^3; $k_B T_e \sim 10$ eV). Therefore, the distribution function of the electrons transferred by the swift ionization from the valence to the conduction band is rightfully described by the Maxwell–Boltzmann distribution.

3.2 Electrons' Collision Rates: Momentum and Energy Transfer

At low excitation (low intensity), electrons are transferred to the conduction band by the threshold-free MPI process. The electrons' number density is low, and the lattice is slightly perturbed and electrons interact mainly with the lattice vibrations (phonons).

The electron-to-lattice interaction can be considered as emission and absorption of phonons. On the other hand, emission of phonon can be considered as a small lattice distortion. More electrons emerge, created by ionization and then continuously heated more the lattice distortions are generated. Thus, in contrast to interaction with metals, the effective rate in the beginning of the excitation of dielectrics is close to zero because the electron number density (and therefore the phonons density) is small. At elevated electrons and ions number density the Coulomb collisions become dominant. Let us consider both processes.

3.2.1 Electron–Phonon Collisions

The general kinetic expression allows explicit calculation of the electron-to-lattice collision rate, $\nu_{\text{e-ph}} \approx n_{\text{ph}} \times \sigma_{\text{at}} \times v_e$; $(n_{\text{ph}} \approx n_e)$. The cross section in the electrons interaction with the lattice vibrations (phonons) can be considered approximately constant, $\sigma_{\text{at}} = \pi \times r_{\text{at}}^2$, because the ions remain cold. The electrons' collision rate in the laser-excited dielectrics grows as a strong function of the laser intensity due to increase of the electron number density and their velocity. The collision rate reaches the natural maximum when the decreasing mean free path of electron equals the inter-atomic separation (Eidmann et al., 2000; Allen, 2000), $\nu_{\max} = v_e \times n_{\text{at}}^{1/3}$. Around the maximum a dielectric converted to plasma where the electron–ion Coulomb collisions prevail. The interpolation between the analytic formulas for the electron–phonon collision rate through the maximum to the plasma domain allows describing the transition (see Eidmann et al., 2000).

Electron–phonon momentum exchange rate along with the electrons number density defines the transient optical properties of the laser-excited dielectrics. Most interesting is the parameter space where the competition between the valence and conduction electrons converts the excited dielectric into the absorbing medium at $\varepsilon_{\text{re}} = 0$. The electrons number density at this state following from the above condition is, $n_{\text{e}} \approx \varepsilon_0 \times n_{\text{cr}}$. Let us present as an example the momentum exchange rate in silica excited by the 800 nm light to the state, where $\varepsilon_{\text{re}} = 0$

(n_{cr} = 1.745 × 10^{21} cm^{-3}; ε_0 = 2.13). Taking $v_e \approx 2 \times 10^8$ cm/s and $\sigma_{at} \approx 3 \times 10^{-16}$ cm^2 one obtains $\nu_{e\text{-ph}} \approx 2 \times 10^{14}$ s^{-1} while $\nu_{max}(SiO_2) \approx 3.75 \times 10^{15}$ s^{-1}.

Maximum momentum exchange rate in the excited dielectric is close to that in a metal. However, the dependence of the effective collision rate on the laser intensity in an excited dielectric is much stronger than that in metals. The presented kinetic scaling allows the semi-quantitative estimate of the effective collision rate in the whole range from the initial sate to the maximum.

3.2.2 Momentum and Energy Transfer Collisions in a Dielectric Converted to Plasma

The dielectric is transferred to the metal-like state when the number density of conductivity electrons approaches to the atomic number density (single electron is stripped off each atom). Growth rate of the electrons number slows down and the Coulomb collisions become dominant. After the first ionization (stripping of one electron per atom) is nearly completed the ionization rate slows down as it occurs near the ablation threshold. At this interaction stage, the electron number density might be considered approximately constant making the momentum transfer rate decreasing with the growth of the electron velocity (Kruer, 1988):

$$\nu_{e-i} \approx 3 \times 10^6 \times \ln\Lambda \times n_e \times Z \times T_{e,\mathrm{eV}}^{-3/2}$$

The arguments for the energy transfer rate from hot electrons to atoms (ions) in metals are applicable to the similar exchange in the laser-excited dielectrics (see Chapter 2). Therefore, we adopt the following formula for the connection the momentum transfer rate to the energy transfer rate:

$$\nu_{\mathrm{eff}}^{\mathrm{en}} = b \cdot \left(\frac{m_e}{M}\right) \nu_{\mathrm{eff}}^{\mathrm{mom}} \tag{3.15}$$

The energy transfer time signifies the beginning of the atomic displacements or even hydrodynamic motion. For example in sapphire excited by laser to the metal-like state, when the momentum transfer rate is close to the maximum value of

$\nu_{max} = 5 \times 10^{15}\ s^{-1}$, the energy transfer time comprises $t_{energy} \approx (M/m) \times (\nu_{max})^{-1} = 7.4 \times 10^{-12}$ s.

3.2.3 Effects of the Electrons' Oscillations in the HF Field on the Collision Rates

At high intensity, when a dielectric is transformed to the plasma state, the energy of the electrons oscillations in the intense HF field becomes comparable to the thermal energy of the electrons. In the classical approximation used for the calculation of ionization ($\Delta_g >> \varepsilon_{osc}$), for calculations of the electrons' collisions rate and absorption, the oscillation energy is usually neglected ($\varepsilon_{th} >> \varepsilon_{osc}$). Studies of the high-intensity HF field interaction with plasma revealed the significant effect of the oscillation energy on the collision rates, on the absorption and the heat conduction and found that it is polarization dependent (Silin, 1965). The approximate account for this effect suggested in Chapter 2 for metals is extended here for the electrons in conduction band of intense laser excited dielectrics.

Interaction of electrons in laser-excited dielectric with the acoustical phonons and with the polar optical phonons was also considered in the whole region of ratio of the thermal electron velocity to the oscillation velocity, $v_{th}/v_{osc} = (2m_e \times \varepsilon_{th})^{1/2} \times \omega/eE$, the Keldysh parameter in calculations of the tunnel and MPI ionization (Vinogradov, 1976). It was found that interaction with the acoustic phonons is affected in the case when the electrons thermal energy is comparable with the oscillations energy in both cases of circular and linear polarization. However, it seems physically unlikely that the atomic vibrations remain unaffected in such a case. Therefore, we take into account the decrease in the collision rate near the transition to the plasma state in the same way as it was done for metals in Chapter 2.

3.3 Transient Permittivity in the Laser-Affected Dielectric

Permittivity of dielectrics at room temperature is dominated by the polarization of the valence electrons. At elevated electric field, small in comparison to the atomic field, the small non-linear effect in polarizability appears, that is the Kerr effect. With the further field increase the valence electrons start to be transferred to the conduction band. Contribution of conduction electrons into permittivity increases while contribution of polarization decreases. These processes are accompanied by the modification of the band gap, appearance of the excited states and finally the band gap collapse after transition to plasma state.

It is known from the measurements in equilibrium that the band gap in the narrow-band-gap dielectrics decreases due to heating or doping because the both processes are accompanied by increase in the number density of electrons in the conduction band. In the most studied case of Silicon when the number density of conduction electrons increased from ~ 10^{14} to 10^{20} cm^{-3} the band gap decreased from 1.12 to 0.541 eV (Lautenschlager et al., 1985). Under the excitation by the pulse with duration shorter than electron–phonon energy transfer the band gap collapse might be associated only with the increase of the conduction electron population. Therefore, the bonded electron contribution into the polarization and corresponding part of permittivity both decrease when the number density of the conduction electrons increases. This is the reason for presenting the permittivity in the laser-excited dielectric as a function of the electron number density in the conduction band.

3.3.1 Non-Linear Contributions to the Permittivity at the Low Intensity ($I \ll I_{at}$)

The electrical displacement vector expresses through the polarization per unit volume, P, permittivity, ε, and polarizability, χ, as follows:

$$D = E + 4\pi P = \varepsilon E \equiv (1 + 4\pi\chi)E$$

Polarization of dielectric in a field weak in comparison to the atomic field, $E \ll E_{at}$, is presented as a series of expansion by the small parameter E/E_{at}:

$$P = \sum_{i=1} P^{(i)} = \sum_{i=1} \chi^{(i)} \times \left(\frac{E}{E_{\text{at}}}\right)^i$$

$\chi^{(i)}$ is i-th order susceptibility. The first-order polarization is estimated as follows:

$$P^{(1)} = \chi^{(1)} E = \frac{d^{(1)}}{V} \frac{E}{E_{\text{at}}} = \frac{a_B^3}{V} E$$

Here, it is implicitly assumed that the dipole moment is large, $d^{(1)} = ea_B = a_B^3 E_{\text{at}}$. The atomic field is $E_{\text{at}} = e/a_B^2$, and a_B stands for the Bohr radius. Thus, the first-order susceptibility reads, $\chi^{(1)} = a_B^3/V \approx n_a a_B^3$. Here, n_a is atomic number density. The second- and third-order susceptibilities are expressed through the first-order one, $\chi^{(2)} = \chi^{(1)} E/E_{\text{at}}$, $\chi^{(3)} = \chi^{(1)} E^2/E_{\text{at}}^2$. Let us express susceptibility through intensity taking the atomic intensity from Chapter 1. Then $\chi^{(3)}$ is expressed as follows:

$$\chi^{(3)} = \chi^{(1)} 0.294 \times 10^{-16} \left(\frac{\text{cm}^2}{\text{W}}\right) \times I \left(\frac{\text{W}}{\text{cm}^2}\right) \tag{3.16}$$

The simple reasoning presented above works semi-quantitatively well even for solids as one can see from the example below.

Example: Let us estimate $\chi^{(1)}$ and $\chi^{(3)}$ for silica ($n_a = 6.6 \times 10^{22}$ cm^{-3}). Then, $\chi^{(1)} = 0.825$ and $\chi^{(3)} = 0.2426 \times 10^{-16} \left(\frac{\text{cm}^2}{\text{W}}\right) \times I \left(\frac{\text{W}}{\text{cm}^2}\right)$ giving the permittivity as follows: $\varepsilon = 1 + 4\pi(\chi^{(1)} + \chi^{(3)}) \equiv \varepsilon_0 + \varepsilon_{\text{nl}}$. Here, $\varepsilon_{\text{nl}} = 3.0.4 \times 10^{-16} \left(\frac{\text{cm}^2}{\text{W}}\right) \times I \left(\frac{\text{W}}{\text{cm}^2}\right)$.

3.3.2 Kerr-Nonlinearity in Silica: Temperature Dependence

Kerr non-linearity in solids originates from the polarizability of the anisotropic molecules. It is obvious that the elevated temperature prevents the dipole moment alignment along the imposed electric field creating polarization. One can present non-linear addition to the temperature-dependent permittivity of silica from the calculations (Il'inskii and Keldysh, 1994):

$$\varepsilon_{\mathrm{nl}} \approx 5.93 \times 10^{-16}\left[\frac{\mathrm{cm}^2}{\mathrm{W}}\right]\left\{\frac{k_B T_{\mathrm{room}}}{k_B T + \varepsilon_{\mathrm{osc}}}\right\} \times I\left(\frac{\mathrm{W}}{\mathrm{cm}^2}\right) \tag{3.17}$$

The energy of molecule due to dipole-field interaction remains small in comparison to the thermal energy. However, the oscillation energy compares to the thermal energy at room temperature at intensity 2.8×10^{11} W/cm^2 (at $\lambda \sim 1$ μm) along with growth of the electron temperature. Therefore, the above formula for the Kerr non-linearity was modified including growing electron temperature and oscillation energy. Hence, one might expect that the Kerr-like non-linearity ceased to exist in silica at intensity above 10^{12} W/cm^2.

3.3.3 Permittivity in the Intense-Laser-Excited Dielectric

Permittivity of the laser-excited dielectric can be approximately presented as a sum of the contribution of the electrons transferred to the conduction band and from the polarization of the valence electrons. At low intensity when electrons' ionization is still negligible the Kerr-like non-linearity represents the first modification of the optical properties under the field action. The transfer of electrons into conduction band is the major process affecting permittivity directly and via effect on the collision rate responsible for absorption and heat conduction. Therefore, it was suggested to present the approximate permittivity as a function of the number density of conduction electrons in the following form (Sokolowski-Tinten and von der Linde, 2000; Gamaly and Rode, 2014):

$$\varepsilon \approx \varepsilon_0 + \varepsilon_{\mathrm{Kerr}} - (\varepsilon_0 - 1)\frac{n_e}{n_a} - \varepsilon_{\mathrm{re}}^{\mathrm{el}} + i\varepsilon_{\mathrm{im}}^{\mathrm{el}} \tag{3.18}$$

Here, the first three terms present the contribution of the valence electrons, while the last two terms are the contribution of the conduction electrons responsible for absorption. The third term for the real part is chosen in such a way that permittivity takes the form of that for plasma when all atoms become singly ionized (one electron is stripped off). The above formula gives correct permittivity values in both limits for the initial state ($n_e << n_{cr}$) and for the metal-like state ($n_e \geq n_{cr}$).

The former contribution has a conventional Drude-like form, including the collision rate calculated above:

$$\varepsilon_{\rm re}^{\rm el} + i\varepsilon_{\rm im}^{\rm el} = \frac{n_e}{n_{\rm cr}\left(1 + \nu_{\rm eff}^2 / \omega^2\right)} + i\frac{n_e}{n_{\rm cr}\left(1 + \nu_{\rm eff}^2 / \omega^2\right)}\nu_{\rm eff}^2 \tag{3.19}$$

There are two interaction regimes emerging in the course of the rising laser intensity during the pulse. First regime corresponds to the low-intensity range where the real part of the permittivity is still positive. The electromagnetic wave propagates through a partly transparent excited media. In the second regime at the elevated intensity, the real part of the permittivity turns to become negative. At the places where $\varepsilon_{\rm re} = 0$ the running wave transforms into the evanescent wave, absorbing the electromagnetic energy in the medium converted into plasma. Note that the permittivity depends mainly on the electron density. Therefore, the temporal and spatial mapping of transient permittivity in the laser-excited dielectric reproduces the pattern of the intensity distribution in the interaction region. The spatially dependent optical properties significantly influence the interaction mode.

At some intensity level, the real part of the permittivity decreases to unity making the medium practically transparent, with the reflectivity close to zero. With the further increase in intensity, the positive polarization of the valence electrons is completely cancelled by the negative contribution of the conduction electrons, turning the real part of the permittivity into zero.

3.3.3.1 Reaching the state $\varepsilon_{\rm re} = 0$

The number density of conduction electrons required to achieve the state $\varepsilon_{\rm re} = 0$, follows from (3.18), $n_e \approx \varepsilon_0 \times n_{\rm cr}$, taking into account that $\nu_{\rm eff} << \omega$, and $\varepsilon_0 >> \varepsilon_{\rm Kerr}$.

For example, in sapphire ($\varepsilon_0 = 2.4$) under the action of 800 nm light (critical density, $n_{\rm cr} = 1.75 \times 10^{21}$ cm^{-3}) the real part of the permittivity turns to zero when the number density of conduction electrons constitutes, $n_e = 4.2 \times 10^{21}$ cm^{-3}, that is ~5% from the atomic density.

In general, the spatial distribution of the excited permittivity is three-dimensional and can be calculated following the above recipe. Thus, during the intense laser/transparent dielectric interaction, the laser-affected medium becomes electrically inhomogeneous: the ultra-short laser pulse at some moment begins interaction with a medium possessing the permittivity gradient where the real part of the permittivity changes from positive to negative values. It appears that the mode of interaction dramatically changes, which we consider in the following section.

3.4 Laser Interaction with the Electrically Inhomogeneous Dielectric

The electromagnetic wave interaction with the electrically inhomogeneous dielectric has significant differences from the interaction with the isotropic medium (Forsterling, 1949; Ginzburg, 1964; Denisov, 1957). The electrical inhomogeneity means that the permittivity varies in space while the mass density remains homogeneous. This is exactly the state created during the interaction of the intense pulse shorter than the time for the energy transfer from excited electrons to ions.

Existence of the singularity in the interaction of the light with inhomogeneous non-absorbing dielectric is the inherent property of the Maxwell equations. It can be seen if the Maxwell equations are converting to the equation for the magnetic field (Landau and et al., 1984):

$$\Delta H - \frac{\varepsilon}{c^2}\frac{\partial^2 H}{\partial t^2} + \frac{1}{\varepsilon}\nabla\varepsilon \times (\nabla \times H) = 0 \tag{3.20}$$

The last term in the above equation might be infinite at zero permittivity if the numerator in this term is non-zero. It was found that that occurs in the interaction of p-polarized light with the inhomogeneous dielectric. The interaction of p-polarized laser with the plasma density gradient was thoroughly studied in laser–plasma interactions (Ginzburg, 1964; Denisov, 1957; Kruer,

1988). In what follows we consider interaction of normally and oblique incident polarized laser light with inhomogeneous dielectric created during the interaction. We show the similarity and difference (plasma wave, length scale, collision rate) with the resonance absorption in plasma with the density gradient.

Consider a plane wave propagating with a wave vector being in y, z plane ($k\{0;\ k_y;\ k_z\}$). In this case, the field does not depend on the x-coordinate. The wave impinges a slab where the real part of permittivity is inhomogeneous in the z-direction:

$$\varepsilon(z) \approx \varepsilon_0 - \frac{n_e}{n_{\mathrm{cr}}} = \varepsilon_0(1 - z/L) \tag{3.21}$$

Here, we assumed the linear profile of the electron density, $n_e/n_{\mathrm{cr}} = z/L$. The electrons density gradients and permittivity gradients coincide as in the plasma case. From Poisson's equation, it follows that for the normal incidence and for the oblique incidence of s-polarized wave, E (E_x; 0;0), there is no divergence in the real-zero permittivity point, $\nabla \cdot \varepsilon \cdot \bar{E} = \varepsilon \cdot \nabla \bar{E} + \frac{d\varepsilon}{dz} \cdot E_z = 0.$ ·

3.4.1 Normal Incidence

Wave parameters are E (E_x, 0,0) ~ exp ($-i\omega t$), k (0, 0, k_z). Wave equation becomes

$$\frac{\partial^2 E_x}{\partial z^2} + \varepsilon(z) k_0^2 E_x = 0. \tag{3.22}$$

The above equation is similar to the Schrodinger equation in quantum mechanics. It can be solved in the quasi-classical approximation (WKB–Wentzel, Kramers, Brillouin) that is equivalent to the geometrical optics (Landau et al., 1984). The solution is as follows:

$$E(z) = \frac{E_0}{\varepsilon^{1/4}} \cdot \exp\left(ik_0 \int_0^z \varepsilon^{1/2}(\xi) d\xi \right);\ k_0 = \omega/c \tag{3.23}$$

This solution is valid under assumption that the permittivity is slowly varying in space, $\frac{d\varepsilon}{dz} << \frac{4\pi\varepsilon^{3/2}}{\lambda}$. It is apparent from the above solution that the field amplitude increases as the wave approaches to the zero permittivity point. However, the geometrical optics approximation becomes invalid in the vicinity of the zero point.

For the linear dependence of the permittivity from (3.23) one gets

$$E(z \leq L) \propto \frac{E_0}{\varepsilon^{1/4}} \exp\left(i\frac{2}{3}\varepsilon_0^{1/2} \cdot k_0 \cdot L\right). \tag{3.24}$$

The solution (3.23) converts to the familiar skin effect in the case of space-independent permittivity putting $\varepsilon^{1/2} = n + i\kappa$ and getting

$$E \propto \exp(-\kappa \cdot k_0 \cdot z) = \exp(-z / l_s).$$

3.4.2 Oblique Incidence

Let us consider the wave incident on a slab at angle θ to the normal with the wave vector, k $(0, k_y, k_z)$ setting in plane $\boldsymbol{y}$, $\boldsymbol{z}$, $k_y = k_0 \cdot \sin\theta$.

3.4.2.1 s-Polarization

The electric vector is perpendicular to the plane of the incidence E $(E_x, 0, 0)$. The wave equation reads

$$\frac{\partial^2 E_x}{\partial z^2} - k_0^2(\varepsilon(z) - \sin^2\theta)E_x = 0. \tag{3.25}$$

However, it is seen that the incident wave becomes partly evanescent and reflected at the point where $\varepsilon(z_1) = \sin^2\theta$. Under suggestion $z_1 = 0$ and $f = k_0^2(\varepsilon(z) - \sin^2\theta) \approx -\alpha z$ in the vicinity of $z = 0$ the Eq. (3.25) reduces to the familiar Airy equation. For the linear permittivity from (3.21), one obtains $\alpha = \left(\frac{df}{dz}\right)_{f=0} = \frac{k_0^2\varepsilon_0}{L}$. One can clearly see the difference with the plasma case. Now we can follow arguments from Kruer (1988). Under assumption $k_0 L \varepsilon_0^{1/2} >> 1$, the asymptotic representation of the Airy function can be used in order to obtain the ratio of the maximum intensity amplitude in the turning point to the incident intensity:

$$\left(\frac{E_{\max}}{E_0}\right)^2 \approx 3.6(\varepsilon_0^{1/2} k_0 L)^{1/3} \tag{3.26}$$

Thus, one can expect about fivefold increase in the intensity near the turning point in the laser-excited dielectric slightly larger than in the similar conditions in plasma.

3.4.2.2 p-Polarization

Let us consider the oblique incidence of a plane wave on a layer where the permittivity along z-coordinate passes through zero ($\varepsilon = -a \times z$). The wave vector lies in the plane of incidence, k $(0, k_x, k_y)$, at the angle θ to the normal to the layer surface. The field components are E $(0, E_x, E_y)$, H $(H_x, 0, 0)$. In this case, it is convenient to consider the equation for the magnetic field (3.20), which takes the following form:

$$\frac{d^2H}{dz^2} - \frac{1}{z}\frac{dH}{dz} - (k_0^2 \cdot a \cdot z + k_x^2) \cdot H = 0 \tag{3.27}$$

The wave reflects at the point where $k_0^2 \cdot \varepsilon - k_x^2 = 0$ and after the reflection point it decays exponentially reaching the zero permittivity point, where the field amplitude has a sharp increase (Forsterling, 1949; Landau et al., 1984). In this case, the component of the electric vector goes parallel to the permittivity gradient. From Poisson's equation, the divergence of the field at the point $\varepsilon = 0$ immediately follows, $\nabla \bar{E} = -\frac{1}{\varepsilon}\frac{d\varepsilon}{dz} \cdot E_z$.

It was found analytically that the solution for the magnetic field near the zero point is the following, $H = H_0\left(1 + \frac{k_x^2 z^2}{2} \ln k_x z\right)$ (Forsterling, 1949; Landau et al., 1984). The electric field expressed through magnetic field from the Maxwell equation:

$$E_z = \frac{ic}{\varepsilon\omega}\frac{\partial H_y}{\partial x} \approx -\frac{ck_x}{\varepsilon\omega} H_0 \tag{3.28}$$

It diverges at $\varepsilon = 0$. However, the energy flux in the direction of the permittivity gradient exists even with the infinitesimally small absorption (Gildenburg, 1963). Indeed the energy absorption rate averaged by the many light periods reads

$$Q = \frac{\omega \varepsilon_{\text{im}}}{8\pi} |E_z|^2 = \frac{\omega \cdot k_x^2}{8\pi k_0^2} \cdot H_0^2 \frac{\varepsilon_{\text{im}}}{\varepsilon^2}. \quad (3.29)$$

Presenting $\frac{\varepsilon_{\text{im}}}{\varepsilon^2} = \frac{\varepsilon_{\text{im}}}{a^2 z^2 + \varepsilon_{\text{im}}^2}$ and taking the limit, $\lim_{\varepsilon_{\text{im}} \to 0} \frac{\varepsilon_{\text{im}}}{a^2 z^2 + \varepsilon_{\text{im}}^2} \frac{\pi \delta(z)}{a}$ the above rate reduces to

$$Q = \frac{\omega \cdot k_x^2 \cdot H_0^2}{8 a k_0^2} \cdot \delta(z) \cdot$$

One obtains the energy flow integrating the above formula over z:

$$\langle S_z \rangle = \frac{\omega \cdot k_y^2}{8 k_0^2 \cdot a} \cdot H_0^2 \cdot \quad (3.30)$$

One can see the direct analogy with the resonant absorption in the ideal plasma. Laser field imposed in the z-direction drives electrons' oscillations entering in resonance with the plasma oscillations and exciting plasma waves. The differences relate to the permittivity gradient, $\frac{d\varepsilon}{dz} \approx \frac{\varepsilon_0}{L}$ and the different electrons number density in the resonance, $n_{\text{res}} = \varepsilon_0 \times n_{\text{cr}}$. The maximum amplitude of the electric field in the resonance following from the dumped oscillator equation reads $E_{\max} = E_0 \times \omega/\nu_{\text{eff}}$, from the Maxwell equations, $E_{\max}/E_0 \approx \varepsilon_0/\varepsilon_{\text{im}}$.

3.4.2.3 Band gap modification and collapse

The band restructuring occurs along with the changes in the electronic landscape. Apparently, the band gap between valence and conduction zone decreases when electrons from the valence zone are transferred into the conduction zone. The decrease in the number of the valence electrons changes the attractive part in the screened pseudo-potential and exchange-correlation effects (Abell, 1985). For 10% electron excitation, the fundamental gap, which becomes indirect as in GaP, vanishes, and the bonding–

antibonding gap at X is reduced by almost one half (Kim et al., 1994)

The dependence of the indirect band gap, for example, in silicon, on the conduction electron density supports the above assumption: the band gap decreases by almost 0.2 eV (from 1.1 to 0.9 eV) due to heating to 1000 K, while the electron number density increases up to 2×10^{18} cm^{-3} (Lautenschlager et al., 1985).

The decrease in the band gap should affect both the multiphoton ionization rate and the impact ionization as well. Keldysh accounted for the ionization from the excited states in the calculations of MPI ionization probability (Keldysh, 1965). The effect of the band gap decrease on the impact ionization rate can be estimated in the classical approximation, $\Delta_g > \hbar\omega$.

3.5 Energy Density Thresholds for Achieving the Major Steps in Dielectric-Plasma Transformation

Let us now define the energy density necessary for achieving three milestones on the way of the transformation from the initial state to plasma: (1) $\varepsilon_{\mathrm{re}} = 0$, when major absorption begins; (2) transfer all valence electrons to the conduction band; (3) full ionization (transition electrons to continuum). To achieve these goals, we adopt the following scenario. First, the energy absorbed per unit volume is proportional to the absorbed fluence and it is contained solely in the free electron component. The electrons' number density; temperature and collision rate for each milestone is defined from the physical conditions describing these states.

For the plane electromagnetic wave normally incident on the semi-infinite medium, the energy absorbed per unit volume per unit time, the boundary conditions accounted for, is expressed in the same general form as it was done for metal in Chapter 1:

$$\langle Q_{\mathrm{abs}} \rangle = \frac{\omega \cdot \varepsilon_{\mathrm{im}}}{8\pi} \frac{4E_0^2}{|1+\varepsilon^{1/2}|^2} = \frac{\omega \cdot \varepsilon_{\mathrm{im}}}{c} \frac{4I_0(t)}{|1+\varepsilon^{1/2}|^2}$$

The energy absorbed to the time t is converted to the electrons internal energy as follows:

$$C_e \cdot n_e \cdot k_B T_e = \int_0^t \langle Q_{\mathrm{abs}} \rangle d\tau \,[\mathrm{energy/cm^3}]$$

The electron temperature can be recovered from this equation taking values for n_e, dielectric permittivity and laser fluence at the moment $\boldsymbol{t}$:

$$C_e \cdot n_e \cdot k_B T_e = \frac{2A \cdot F(t)}{l_s}\left[\frac{\mathrm{Joules}}{\mathrm{cm^3}}\right];\ F(t) = \int_0^t I_0(\tau) d\tau$$

We consider the free electrons as the ideal gas with $C_e = 3/2k_B$. Let us define the three thresholds for sapphire under the action of 800 nm laser light ($\omega = 2.356 \times 10^{15}\ \mathrm{s^{-1}}$; $n_{\mathrm{cr}} = 1.75 \times 10^{21}\ \mathrm{cm^{-3}}$).

3.5.1 Threshold to Achieve the State Where $\varepsilon_{\mathrm{re}} = 0$

The number density of conduction electrons transferred through the band gap of $\Delta_{\mathrm{Sap}} = 9.8$ eV follows from condition, $\varepsilon_{\mathrm{re}} = 0$, it equals $n_e \approx \varepsilon_0 \times n_{\mathrm{cr}}$. The Coulomb collision rate for the momentum exchange is $\nu_{\mathrm{e\text{-}i}} \approx 1.1 \times 10^{15}\ \mathrm{s^{-1}}$. One immediately calculates the imaginary part of the permittivity, $\varepsilon_{\mathrm{im}} \approx \varepsilon_0 \times (\nu_{\mathrm{eff}}/\omega) = 1.40$; refractive index, $n = \kappa = 0.837$; absorption coefficient, $A = 0.822$. The absorbed laser fluence required for the formation of the zero permittivity state ($\varepsilon_{\mathrm{re}} = 0$) can be found from the condition that all absorbed energy density confined solely in the thermal energy of electrons transferred to the conduction band. Then the threshold fluence is defined by the following equation:

$$F(t_{\varepsilon=0}) = \frac{3 \cdot \Delta_g \cdot n_{\varepsilon_{\mathrm{re}}=0} \cdot c}{4A \cdot \omega \cdot \kappa} \tag{3.31}$$

Then, $F(\varepsilon_{\mathrm{re}} = 0) = 0.25\ \mathrm{J/cm^2}$.

3.5.2 Threshold Fluence for Transferring All Valence Electrons to the Conduction Band

In these conditions $n_e = n_a = 1.18 \times 10^{23}$ cm^{-3}; $\nu_{\text{eff}} = \nu_{\max} = 5 \times 10^{15}$ s^{-1}; $A = 0.33$; $\kappa = 4.45$. Then from (3.31) follows the threshold for transferring dielectric into metal-like state $F_{d\text{-}m} = 1.2$ J/cm^2.

3.5.3 Ionization Threshold: Transfer Electrons to the Solid Plasma State

In this case, electron shall gain energy in excess of ionization potential. Ionization potentials for Al and oxygen are, respectively, $J_{\text{Al}} = 5.986$ eV; $J_O = 13.6$ eV. One may expect the higher threshold for transition to plasma state. Indeed, the damage or ablation threshold for sapphire from experiments is around 2–2.5 J/cm^2.

3.6 Energy Equations: Two-Temperature Approximation

The distribution function in the form close to that of the equilibrium Boltzmann–Maxwell establishes in the electronic and lattice (ions) sub-systems faster than the electron–ion energy equilibration. We shall remind that this is the distribution function still bearing non-equilibrium features—it lacks the high-energy tail. It needs a few tens of picoseconds to establish the full equilibrium distribution. Therefore, for considering the material transformation similar (but not identical) to the equilibrium counterparts like melting and evaporation the exact distribution function should be considered. The processes of the energy re-distribution and balance as a rule are described in so-called two-temperature (2T) approximation. It is assumed that the electronic ($\boldsymbol{T}_e$) and lattice ($\boldsymbol{T}_L$) temperature (in the sense of the average energy per particle) are quickly established (Kaganov et al., 1957, Anisimov et al., 1974). Two coupled energy equations for electrons and ions are the following:

$$\frac{\partial E_e}{\partial t} = Q_{\text{abs}} - C_e \cdot n_e \cdot k_B (T_e - T_L) \nu_{\text{en}} - Q_{\text{heat}} \tag{3.32}$$

$$Q_{\text{abs}} = \frac{\omega}{c} \varepsilon_{\text{im}} \cdot I(t)$$

$$\frac{\partial E_L}{\partial t} = C_e n_e \cdot k_B (T_e - T_L) \nu_{\text{en}}$$

Here, in the first equation for the energy of electrons, the first (source) term is the energy absorbed by electrons per unit time and unit volume. The second term is the energy sink into the lattice sub-system (that is the source term in the lattice temperature equation). The third term in the first equation describes the losses due to heat conduction, which we discuss in the next paragraph, $Q_{\text{heat}} = \nabla \bar{q}_s$.

3.7 Linear and Non-Linear Heat Conduction: Transition from Heat Transfer in Solid to That in Plasma

There is no energy flow in a medium where the temperature, pressure, density, velocity are homogeneous. In the non-dissipative medium the energy flux of the mono-energetic electrons expresses as $\bar{q} = \varepsilon_e n_e v_e$ [Joule/cm^{2}s]. The general form of the energy flow in a medium where the electronic distribution function, f_e, is known has a form, $q = \int \varepsilon_e v_e f_e d^3 v_e$. In the isotropic medium where the only temperature is inhomogeneous the heat flow is proportional to the temperature gradient $\bar{q}_\alpha = -\kappa_{\alpha\beta} \frac{\partial T}{\partial x_\beta}$. Note that the heat conduction coefficient is a tensor in thermally anisotropic medium.

3.7.1 Heat Diffusion in the Isotropic Medium

Energy is transported by the flow of the energy carriers experiencing collisions with the different particles in a medium. The diffusion approximation for description of this flow is applicable when mean free path for the single collision is much less than the scales of the temperature gradient or that for the heated area. Heat flux transported by particles with the heat capacity, C, number density, n, in the isotropic medium in the diffusion approximation has a form

$$\bar{q}_{\alpha} = -\kappa_{\text{heat}} \frac{\partial T}{\partial x_{\alpha}} \equiv -C \cdot n \cdot D \frac{\partial T}{\partial x_{\alpha}}. \tag{3.33}$$

Here, D is the diffusion coefficient expressed through the characteristic velocity of the energy carriers, v, and mean free path for collisions, l_{mfp}. The last one can be expressed through the particle's velocity and collision rate, $l_{\text{mfp}} = v/\nu_{\text{eff}}$. Thus, the diffusion coefficient is

$$D = \frac{l_{\text{mfp}} v}{3} = \frac{v^2}{3\nu_{\text{eff}}} [\text{cm}^2/\text{sec}]. \tag{3.34}$$

Conventional heat conduction coefficient from (3.33) reads

$$\kappa_{\text{heat}} \equiv -C \cdot n \cdot D\, [\text{W/K} \times \text{cm}].$$

One can estimate the cooling time for the sample with the size, L, through the diffusion coefficient as follows:

$$t_{\text{cool}} \approx \frac{L^2}{D} \tag{3.35}$$

Equally diffusion coefficient allows estimating distance travelled by the heat front, $x_{\text{front}} \approx (Dt)^{1/2}$.

3.7.2 Thermal Conduction in Dielectrics: Energy Carriers are Phonons

In dielectrics at the temperature below the Debye temperature, the heat carriers are phonons. The phonons, considered as quasi-particles, have the effective collision rate proportional to the temperature, $\nu_{\text{ef,ph}} \approx \frac{k_B T}{\hbar}$, while the velocity of phonons is $v_{\text{ph}}^2 = \frac{3k_B T}{M}$, giving the diffusion and heat conduction coefficient temperature independent, $D_{\text{ph}} = \frac{v_{\text{ph}}^2}{3\nu_{\text{ph}}^{\text{mom}}} = \frac{\hbar}{M}$. For example, in the dielectric with atom's mass, $M = 20$ a.u., one gets $D \approx 0.3 \times 10^{-4}$

cm^2/s. Thus, for the micron size sample, $L = 10^{-4}$ cm, the cooling time is $t_{cool} \approx 3 \times 10^{-4}$ s.

3.7.3 Electronic Heat Conduction in Plasma

In isotropic ideal plasma created by the laser pulse shorter than the energy transfer time from hot electrons to ions ($T_e >> T_{ion}$) the Coulomb collisions are the dominant processes. The electron–electron collisions establish the quasi-equilibrium energy distribution in the electron sub-system during the time inversely proportional to the electron plasma frequency, $t_{ee} \approx (\omega_{pe})^{-1} \sim 0.1$ fs. Thus, one can use the collision rates for the momentum exchange as in simple plasma with equal number of electrons and singly ionized ions (Kruer, 1988):

$$\nu_{ei} \approx 3 \cdot 10^{-6}\, ln\Lambda \cdot \frac{n_e}{(k_B T_e / eV)^{3/2}} [s^{-1}] \tag{3.36}$$

Electron number density is in cm^{-3}. Diffusion coefficient in such plasma is inversely proportional to the electron density and strongly dependent on the electron temperature:

$$D_{ei} = \frac{v_e^2}{3\nu_{ei}} \propto \frac{1}{n_e}\left(\frac{k_B T_e}{eV}\right)^{5/2} \tag{3.37}$$

The heat conduction coefficient is expressed as

$$\kappa_e \equiv C_e \cdot n_e \cdot D_{el} \propto \left(\frac{k_B T_e}{eV}\right)^{5/2}. \tag{3.38}$$

The heat conduction coefficient in homogeneous plasma becomes temperature dependent and density independent. Therefore, the heat transfer process becomes non-linear in respect to the electron temperature. Dielectric converted to the plasma state at the electron temperature approximately equal to $k_B \times T_e \sim 10$ eV ($v_e \approx 2 \times 10^8$ cm/s) when the collision rate is close to the maximum, $\nu_{max} \approx 5 \times 10^{15}\ s^{-1}$. Therefore, in plasma state $D \approx cm^2/s$ and heat conduction coefficient is close to that as in good metals.

It is instructive to check the applicability of the diffusion approximation, $l_{\text{mfp}} << L$, which is the basis of heat conduction equation.

3.7.4 Applicability of Diffusion Approximation for Description the Heat Conduction in the Laser-Created Plasma

In the absence of losses (collisions in the considered case), the free energy flow of electrons, carrying energy density, $E_e = C_e \cdot n_e \cdot k_B T_e$, expresses as $q_{\text{free}} = E_e \cdot v_e$ [W/cm^2]. Let us present the heat flow of (3.33) in an approximate form allowing seeing the difference of the diffusion heat flow from the free flow:

$$q_{\text{heat}} \approx -E_e \cdot v_e \frac{l_{\text{mfp}}}{\Delta x_T}$$

Here, the scale of the temperature gradient is explicitly introduced, $\Delta x_T \approx \left(\frac{1}{T_e}\frac{\partial T_e}{\partial x}\right)^{-1}$. Thus, at the diffusion approximation, which is valid at $\Delta_x >> l_{\text{mfp}}$, the heat flux is much less than the free energy flow. Let us now check the applicability of the diffusion approximation in the dielectric converted to the plasma at $k_B \times T_e \sim 10$ eV ($v_e \approx 2 \times 10^8$ cm/s) when the collision rate is, $\nu_{\max} \approx 5 \times 10^{15}$ s^{-1}. Thus, mean free path is $l_{\text{mfp}} = v_e/\nu_{\max} \approx 4 \times 10^{-8}$ cm. In these conditions the absorption length is around several tens of nanometres, $l_{\text{abs}} \approx (5\text{–}6) \times 10^{-6}$ cm $>> l_{\text{mfp}}$, making legitimate the use of the diffusion approximation for the heat conduction description. The electron mean free path in plasma grows as the electron's energy increases as $l_{\text{mfp}} \sim (k_B \times T_e)^2$. Hence, at $k_B \times T_e \approx 100$ eV, the mean free path becomes comparable to the absorption length that is an approximation for the inverse temperature gradient. The extremely intense pulses with intensity in excess of 10^{15} W/cm^2 could deliver such electron energy, which triggers ablation. Another possible reason for the violation of the diffusive heat transport might be the generation of the electrons from the high-energy tail of the spectrum. However, under dielectric excitation by ultra-short (femtosecond) laser pulses, the electrons' and ions' distribution functions are deficient in the high-energy tail.

Note that in HF electric field at intensity exceeding 10^{14} W/cm^2 the oscillation energy becomes comparable with the electron's thermal energy for the laser wavelength around one micron. This effect, discussed above, leads to the decrease in the collision frequency and therefore for increase in the mean free path.

3.7.5 Linear and Non-Linear Heat Propagation from the Focal Region

It is instructive first consider a simple case where the heat propagation can be treated in one-dimensional approximation. Let us analyse the laser interac tion with a flat target at the normal incidence. Conventionally the focal spot size is a few microns, while the absorption depth reaches a few tens of nanometres, i.e., two orders of magnitude smaller. Therefore, it is reasonable to describe the heat propagation as one-dimensional (1D) in the direction normal to the target surface. The heat conduction equation is the electrons' energy equation of (3.32) after the energy source is switched off and the only losses are due to escape of the energy carriers. It describes electron temperature decrease in space and time:

$$\frac{\partial E_e}{\partial t} = -Q_{\text{heat}} = -\frac{\partial q}{\partial x} \equiv -\frac{\partial}{\partial x}\kappa_e \frac{\partial k_B T_e}{\partial x}$$

It reduces to the familiar form of the heat conduction equation in the homogeneous medium where the heat capacity and number density is constant:

$$\frac{\partial T}{\partial t} = \frac{\partial}{\partial x} D \frac{\partial T}{\partial x} \tag{3.39}$$

The energy per unit area (fluence) delivered to the surface of semi-infinite slab in the 1D case with the help of the energy conservation law expresses as follows:

$$\int_0^{x_f} T_e dx = \frac{F}{n_e \cdot C_e} = Q. \tag{3.40}$$

3.7.5.1 Linear heat conduction

The diffusion coefficient is temperature independent at the moderately elevated temperature. This is a case of the linear heat conduction. Equation (3.39) has the well-known solution (Landau et al., 1984):

$$T = \frac{Q}{(4\pi \cdot Dt)^{1/2}} \exp\{-x^2/4Dt\} \tag{3.41}$$

Formally, the temperature in the linear heat conduction process described by (3.41) reaches infinity instantaneously (being infinitesimally small). The heat front propagation length estimated as, $x_{\text{fr}} \approx (D \cdot t)^{1/2}$, agrees with the experiments. This space/time temperature dependence is useful for the temperature estimation when the temperature rise due to laser excitation is moderate and ionization degree is low.

3.7.5.2 Non-linear heat propagation

In plasma, the heat diffusion coefficient is temperature dependent, $D = aT^m$ (see (3.37). The heat propagation front is steep. The space temperature dependence is the following (Zel'dovich and Raizer, 2002):

$$T = \left[\frac{mv_f}{a} |x_f - x| \right]^{1/m} \tag{3.42}$$

Here, v_f and x_f are, respectively, the velocity and coordinate of the heat front. Using heat conduction Eq. (3.39) and energy conservation (3.40), one can develop the approximate scaling useful for the practical estimates. Indeed, from (3.39) follows:

$$x_f^2 \approx Dt \approx aT^m t. \tag{3.43}$$

From the energy conservation, follows dependence on the absorbed fluence, F:

$$T \approx Q/x_f \propto F \tag{3.44}$$

Combination of (3.43) and (3.44) produces dependence of the heat front coordinate on time and absorbed fluence:

$$x_f \approx (aQ^m t)^{1/(m+2)} \propto F^{5/9} t^{2/9} \tag{3.45}$$

Here, we put m = 5/2 as for plasma. We are considering the electronic heat conduction in non-equilibrium plasma where the electrons are swiftly heated to the high temperature while the ions remain cold. Hence, the non-linear heat wave front decelerates quickly due to energy transfer from hot electrons to cold ions:

$$v_f \approx \frac{x_f}{t} \approx (aQ^m)^{1/(m+2)} t^{-\frac{(m+1)}{(m+2)}} \propto F^{5/9} t^{-7/9} \tag{3.46}$$

The scheme of the non-linear heat wave propagation is presented in Fig. 3.1.

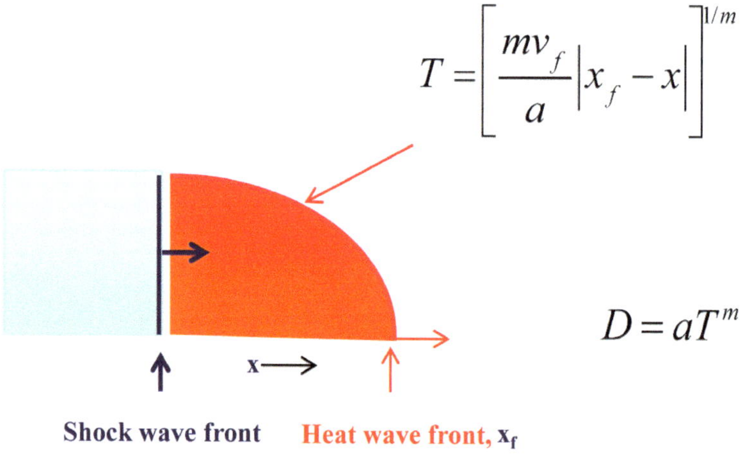

Figure 3.1 The non-linear heat wave propagation in plasma.

Initially heat wave velocity is super-sonic. The non-linear heat wave stops when the heat front velocity becomes equal to the local speed of sound. At this stage the electron and ion temperature equilibrates, ions starts to move generating the shock wave.

3.8 Summary

- *Tunnelling* effect dominates the ionization of the wide-band gap dielectrics at the intensity $\geq 10^{14}$ W/cm^2.
- *Probability of tunnelling* by the circular-polarized light is much larger than that for the linear polarization
- Probability of MPI *by the circular polarized light* (absorption of photons with spin one) *is less* than that by absorption of photons of zero spin.
- Ionization by the electron impact overwhelms the MPI for intensities in a range 10^{13}–10^{14} W/cm^2; the avalanche mode dominates at higher intensity.
- The short-laser-excited dielectric is electrically inhomogeneous (permittivity maps intensity distribution).
- The amplitude of the resonantly increased field at the zero permittivity point in laser-excited dielectric is higher, than in plasma resonance.
- Band gap shrinks and collapses when around 10% of valence electrons are promoted into the conduction band.
- Following are the examples of laser-induced threshold fluences for transformations in sapphire:
  - $F(\varepsilon_{\text{re}} = 0) = 0.25$ J/cm^2; ~ 10% of electrons transferred from the valence to the conduction band,
  - $F_{d\text{-}m} = 1.2$ J/cm^2; full single ionization fluence, 2–2.5 J/cm^2

Chapter 4

Non-Destructive Transformations: Formation, Lifetime and Decay of Unconventional States of Matter

Swift excitation of solids by ultra-short intense laser pulses transforms any material to a transient state where electrons may possess high energy while the atoms remain in the initial state. Laser-excited solid holds on a non-equilibrium state until the energy transfer from electron to the lattice and temperature equilibration. Therefore, the question arises how a solid may be re-structured in these non-equilibrium conditions. In this chapter, we consider the ultra-fast transformations induced by the short intense laser at the absorbed energy density below the damage threshold while the laser-affected material remains intact.

Heating and applied pressure in equilibrium may lead to crystal-to-crystal transformation. Ultimate disordering of a solid can be achieved by heating over the melting point (conversion to a liquid) accompanied by changes in the mass density. Many experiments were performed in order to find the differences between the melting in equilibrium and its non-equilibrium counterpart—swift raise the energy density over the equilibrium enthalpy of melting. In what follows, we describe the excitation

Femtosecond Laser-Matter Interactions: Solid-Plasma-Solid Transformations at the Extreme Energy Density
Eugene Gamaly

ISBN 978-981-4877-40-4 (Hardcover), 978-1-003-25661-8 (eBook)
www.jennystanford.com

of different materials (metals, semi-metals and dielectrics) by short intense pulses during the period before the temperature equilibration. It is instructive to consider all consecutive stages of electronic excitation during the pulse and after the end of the pulse and identify the forces displacing the atoms and finally disordering the initial structure. It was demonstrated in numerous experiments that laser-excited material transforms into the short-lived state different from the conventional solid or liquid. The material electronic and spatial re-structuring results in the changes of optical properties, which can be detected by the external probe.

It was observed that the optical probe beam reflected from the laser-excited solid oscillates with the frequency of the cold phonons in this material (Cheng et al., 1991; Zeiger et al., 1992). Further studies demonstrated that generation of the coherent phonons came first followed by the decay of the harmonic modes and instability eventually leading to the lattice disorder and formation of the short-lived state different from solid and conventional melt.

Swift excitation of the transparent dielectrics by the ultra-short intense laser pulse leads to the electron transfer from the valence band to the conduction band generating many transient material states. Laser re-shaping of the electronic landscape results in the generation of excited states inside the band gap finally leading to the gap shrinkage.

It is important to find the new phase states, which could be created by short intense pulse and then preserved in the meta-stable state well after the end of the pulse. Such states were generated in photo-refractive materials, which are not considered here. We describe below the formation and properties of the short-lived states in a hope that understanding of their nature will allow finding the ways to make them meta-stable.

4.1 Generation and Decay of the Coherent Phonons

We recollect the experiments and the theory describing the generation and decay of the cold coherent phonons as an initial step finally leading to the laser induced material disordering.

We consider the processes in the laser-excited solid at the absorbed energy density close to the equilibrium enthalpy of melting (fusion). Equilibrium enthalpy of fusion for majority of solids, including metals and dielectrics, is in a range of H_f = (1–0.3) kJ/cm^3 (see the table in Appendix 3). The fluence of (2–0.6) mJ/cm^2 absorbed in a layer of 40 nm thick generates such energy density. The 100 fs laser pulse with the average intensity 10^{10} W/cm^2 generates the incident fluence of mJ/cm^2. Note that in order to deposit the energy density sufficient to melt a dielectric one needs the intensity and fluence approximately two-three orders of magnitude higher than for metals making a dielectric ionized and absorptive. The ionization threshold for dielectrics is around 0.2 J/cm^2. Below we consider the laser interaction mainly with metals and semi-metals at intensity of 10^9–10^{10} W/cm^2 where the absorbed energy density is sufficient for the electron excitation. The optical properties and relaxation processes are treated explicitly and compared to the experimental results.

4.1.1 Melting in Equilibrium: Atomic Vibrations Become Anharmonic

We start from the description of the ultra-fast laser-excitation of semi-metal Bi. The properties of this semi-metal were thoroughly studied in equilibrium and in the non-equilibrium conditions theoretically and experimentally in the numerous laboratories around the world. These studies allow comparing the material transformations in the different conditions and identify the differences. Let us first recollect the sequence of the processes leading to melting of Bi in equilibrium. Enthalpy of fusion for Bi is 0.5 kJ/cm^3 (Landolt-Börnstein, 1983), just in the middle of the whole range of the melting enthalpy for different solids. The temperature-dependent dielectric permittivity of Bi was measured in equilibrium in the temperature range from room temperature up to 200 K over the melting point of 544.7 K (Comins, 1972; Garl, 2008). At the room temperature, atoms perform harmonic vibrations around their equilibrium positions in the lattice. The atomic motion loses its oscillatory character and becomes chaotic (randomized) when the change in the average atomic position constitutes significant part of the oscillation amplitude. The atomic shift due to non-linear interaction of phonons calculated at the melting point coincides with the average displacement following from the Lindemann criterion of melting (Lindemann, 1910; Pines, 1964). It is known that the formation of the thermal defects (7–10% from the total number of atoms) is crucial for the solid disordering (Fecht, 1992). These defects (vacancies) are generated by the atoms with the energy exceeding $\Delta H_{\text{vac}} = 9.28 \times k_B T_{\text{melt}}$ from the energetic tail of the Maxwell distribution that continuously adjusts to the growing temperature during the equilibrium heating.

The measured dielectric permittivity of bismuth near and over the melting point was described well by the Drude-like formula (Comins, 1972; Garl, 2008; Gamaly, 2011). From these measurements, the real and the imaginary parts of the permittivity along with the effective momentum transfer rate were recovered: at room temperature $\varepsilon_{\text{re}} = -16.25$, $\varepsilon_{\text{im}} = 15.4$, $\nu_{\text{eff}} = 2.1 \times 10^{15}\ \text{s}^{-1}$; at $T = 773$ K; $\varepsilon_{\text{re}} = -11$; $\varepsilon_{\text{im}} = 28.9$, $\nu_{\text{eff}} = 5.67 \times 10^{15}\ \text{s}^{-1}$. The experimental data are well described by the Drude formulae under assumption that over the melting point all five valence electrons were transferred to the conduction band (at room temperature there were only three conduction electrons). Effective collision rate increases in proportion to the temperature (as electron–phonon collision rate in a sold does) up to the temperature 200 K higher than the melting point. Thus, the electrons' interaction with atoms within a melt might be considered as small atomic displacements when the electron–lattice interactions treated in the deformation potential approximation as a lattice distortion (Kittel, 1996). We are keeping the term "electron–phonon" effective collisions even for a melt and in the non-equilibrium transformations later in the book where the phonon notion becomes questionable.

One can see that momentum and energy exchange rates between electrons and lattice are much faster than the heating rate in equilibrium. Therefore, using adiabatic approximation and Virial theorem (equality of the oscillation energy to the potential energy in the harmonic vibration), one can estimate the frequency and amplitude of the phonons (Il'inskii and Keldysh, 1994), which one can use in the non-equilibrium conditions considered later. Potential energy of harmonic vibrations expresses as the first (quadratic) term in expansion of the inter-atomic potential in the powers of the atomic displacement (phonon's amplitude). Equalizing the potential and kinetic energy of vibrations allows estimation of the phonon's frequency from the following:

$$\Delta U_e = \frac{1}{2}\left(\frac{\partial^2 U}{\partial q^2}\right)_0 \cdot q^2 \approx M\omega_{\text{ph}}^2 \cdot q^2/2 \tag{4.1}$$

We estimate $\left(\frac{\partial^2 U}{\partial q^2}\right)_0 \approx \varepsilon_b/d^2$. Here, q is the cold phonon amplitude, ε_b is the binding (cohesive) energy, and d is the inter-atomic distance in equilibrium (see also Chapter 2). The cold phonon frequency is then expressed as follows:

$$\omega_{0,\text{ph}}^2 \approx \varepsilon_b/M_{\text{Bi}} \times d^2 \tag{4.2}$$

For Bi (ε_b = 2.16 eV, d = 3.3 Å, M = 3.67 × 10^{-22} g) one gets ω_{ph} = 3 × 10^{12} s^{-1}. It was implicitly suggested in the above estimates that the centre of gravity of the unit cell is in the rest. Therefore, the above formula gives the frequency close to that of the optical phonon, of the A_{1g} mode, $\omega_{\text{ph}} \approx \omega_{\text{A1g}}$ (oscillation along the c-direction). The cold phonon's amplitude estimates using the uncertainty relation and the above formulae as follows:

$$q_{0,\text{ph}} \approx \left(2\hbar/M_{\text{Bi}}\omega_{0,\text{ph}}\right)^{1/2} \tag{4.3}$$

Using Bi parameters one gets the cold phonon's amplitude around 0.2 Å close to the prediction of the Lindemann criterion for the atomic displacement at the temperature approaching the melting point.

Summing up one concludes that heating in equilibrium up to and above the melting point results in transfer of all valence electrons to the conduction band along with strong enhancement of the electron–phonon interaction (in a sense of hot electron-induced atomic displacements). Moreover, using only the electronic component one can describe the changes in the electronic properties with the reasonable accuracy taking into account experimentally observed linear dependence of the electrons effective collision rate on the temperature.

However, it is obvious that the equation of state, describing the heated Bi, changes. Estimate of the melt temperature in Bi taking the Dulong–Petit heat capacity and treating melt as the ideal gas underestimates the melting point on 20%. Description of the heated electrons as the degenerated electron gas seems adequate and it holds to the electron temperatures well over the equilibrium melting point.

4.1.2 Swift Deposition of the Energy Density Comparable to the Enthalpy of Fusion by the Ultra-Short Laser Pulse

We consider the intense sub-picosecond laser delivering the energy density comparable and exceeding the equilibrium enthalpy of fusion during the time much shorter than the time for the energy transfer from swiftly heated electrons to atoms.

In the experiments (Boschetto et al., 2008; Garl et al., 2008) laser pulse (40 fs, 800 nm, F = (2.7–6.7 mJ/cm^2) excited the single crystal of Bi (1,1,1) oriented. The time-dependent reflectivity from the excited target was measured with the weak optical probe of the same pulse duration with the relative accuracy of 10^{-5}. Time-dependent optical probe reflection from the laser-excited Bi was found oscillating with the frequency close to that of A_{1g} phonon's mode in the cold Bi.

The absorbed energy density was in a range E_{abs} = (0.64–1.59) J/cm^3. It should be compared to 0.5 J/cm^3 of the equilibrium enthalpy of fusion. We suggest on the basis of the equilibrium measurements that the strong rapid heating transfers all valence electrons into the conduction band and take the momentum exchange rate the same as extracted from the equilibrium experiments. That allowed defining the permittivity, absorption coefficient, A, and absorption depth, l_{abs}. The absorbed energy density is converted to the internal energy of the electrons sub-system. The maximum electron temperature reached at the end of the pulse is then obtained using the total laser fluence, *F*:

$$E_{\text{abs}} = \frac{A \cdot F}{l_{\text{abs}}} = C_e \cdot n_e \cdot T_e = \frac{\pi^2}{2\varepsilon_F} \cdot n_e \cdot k_B^2 T_e^2 \tag{4.4}$$

In the experiments (Garl et al., 2008), the fluence range F = (2.7–6.7 mJ/cm^2) corresponds to the absorbed energy density E_{abs} = (0.64–1.59) J/cm^3 and the maximum electron temperature is in the range of T_e = 1,514–3,761 K ($k_B T_e$ = 0.325–0.13 eV). During the pulse, the laser electric field may affect material state via direct field action (polarization and field pressure) along with the effect of absorption and electrons' heating. Therefore, it is instructive to consider the processes during the pulse and after the end of the pulse separately.

4.1.2.1 Processes during the pulse: building the electronic pressure force

Electrons are heated to end of the pulse to the temperature three times larger than in equilibrium at the absorbed energy density equal to the enthalpy of fusion. Moreover, the absorbed energy density exponentially decreases inside the laser-excited layer creating the pressure gradient and correspondingly the force, $F_{\text{th}} = \nabla P_e / n_e$. Average absorbed energy density kJ/ cm^3 corresponds to the GPa pressure much less than the strength of the crystal assuring that laser excitation not damaging the sample. Estimate of the electronic pressure gradient force is straightforward assuming the inverse gradient scale equals to the absorption length:

$$F_{\text{th}} = \nabla P_e / n_e \approx E_{\text{abs}} / l_s \cdot n_e \tag{4.5}$$

For the parameters of the experiments (Garl et al., 2008), where the absorbed energy density varied in the range of (1–3) of the equilibrium enthalpy of melting, the electronic force is in the range, $F_{\text{th}} \approx (1.7\text{–}4) \times 10^{-15}$ J/cm. During the pulse time, the field pressure force (light pressure) and polarization force are also active. However, these two forces are much smaller than the electronic pressure force (see for more details in (Gamaly, 2011). Note also that the pressure force in turn is much smaller than the elastic force driving the atomic vibrations. Therefore, this force cannot change the cold phonon's frequency. Moreover, this force has a maximum value at the end of the pulse, which is much shorter than the phonon's period ($2\pi/\omega_{\text{ph}} \approx 2$ ps for A_{1g} mode in Bi). Thus, this force acts as an almost instantaneous weak blow on the harmonic oscillator vibrating with the cold phonon's frequency. The elastic force is slow one and it is not effective during the time much shorter than the period of atomic vibration. On the long time scale after the end of the laser pulse, the laser-imposed force acts as a small perturbation:

$$q_k(t) = q_{0,k}(t) + \Delta q_k(t) \tag{4.6}$$

It is reasonable to assume that the perturbation oscillates with the same frequency as unperturbed vibration and the laser force affects only the vibration amplitude. Thus, the change in the phonon's amplitude induced by the laser-exerted force in the k-direction takes the following form:

$$\Delta q_k(t) = (F_{\text{th},k}/2\omega_{\text{ph}} \cdot \gamma \cdot M)\exp\left\{(\omega_{\text{ph}}^2 - \gamma^2)^{1/2} t - \varphi\right\} \tag{4.7}$$

4.1.2.2 After the end of the pulse: electronic heat transfer, electron–lattice temperature equilibration, and phonon's decay rate

Initially purely harmonic motion gradually attains non-linear character. Amplitude and frequency of vibrations decrease due to lattice heating during the energy transfer from electrons to the lattice. The interaction between different phonon modes becomes significant finally dumping the oscillations. The damping coefficient γ taking into account these processes is introduced in (4.7). The effective collision rate retrieved from the temperature dependent reflection measurement in equilibrium, $\nu_{\text{eff}} = 5.67 \times 10^{15}\ \text{s}^{-1}$ (Comins, 1972; Gamaly, 2011) at the temperature above the melting point allows estimate effects of cooling due to the electronic heat transfer and electrons–lattice temperature equilibration. The electronic heat diffusion coefficient is estimated as $D = v_F^2/3v_{\text{eff}}$ = 2.35 cm^2/s. It is much larger than that in equilibrium D_{eq} = 0.067 cm^2/s (Landolt-Börnstein, 1983). However, heat diffusion extracted from experiments with the ultra-fast electronic probe (Johnson et al., 2008) gave $D \approx 2$ cm^2/s. Hence, the cooling of the absorption layer of 27.8 nm appears to be fast, $t_{\text{cool}} = (l_s)^2/D = 3.8$ ps. In contrast, the electrons–lattice temperature equilibration appears longer than the observation time, $t_{e-L} = \left(\frac{m_e}{M} v_{\text{eff}}\right)^{-1} \approx 71$ ps. Therefore, supposedly the phonon–phonon interaction might be the major reason for the oscillation damping. Probability of decay of an optical phonon into two acoustic phonons per unit time can be calculated with the help of the quantum perturbation theory as third-order term (Il'inskii and Keldysh, 1994). The inter-atomic potential perturbed by the swift weak laser action can be presented as expansion into series of the powers of atomic displacement:

$$U = U_0 + \frac{1}{2}\left(\frac{\partial^2 U}{\partial x^2}\right)_0 \cdot q^2 - \frac{1}{6}\left(\frac{\partial^3 U}{\partial x^3}\right)_0 \cdot q^3 + \cdots \tag{4.8}$$

The first term in this expansion associated with the phonon's energy allows estimating the cold phonon frequency, close to that of A_{1g} mode as it was demonstrated above (see Eq. 4.2 and the following). The perturbation Hamiltonian responsible for the phonon's decay can be qualitatively estimated as the third-order term in the above expansion (4.8), $H_{\text{pert}} \approx \frac{(3k_BT/2)^{3/2}}{6\varepsilon_b^{1/2}}$. The probability per unit time of the optical phonon's decay into two acoustic phonons expresses through the perturbation Hamiltonian (Il'inskii and Keldysh, 1996):

$$\omega_{\text{decay}} \approx 2\pi\,|H_{\text{pert}}|^2/\hbar \cdot k_BT \propto (k_BT)^2/\hbar \cdot \varepsilon_b \tag{4.9}$$

One can see that for Bi parameters, the damping coefficient in (4.7) is estimated as $\gamma \approx w_{\text{decay}}$, and it is in a range of few picoseconds. That is in a qualitative agreement with the experiments of (Boschetto et al., 2008; Garl et al., 2008).

The swift excitation of electrons decreases the attraction part of the inter-atomic potential correspondingly slightly decreasing the binding energy, $\varepsilon_b \approx \varepsilon_{0,b} - k_B T_e$, increasing the inter-atomic distance and softening the phonon's mode while the lattice remains cold. The condition, $k_B T_e << \varepsilon_b$, holds in the experiments (Boschetto et al., 2008; Garl et al., 2008). The estimation of the effect of the electrons' temperature on the phonon's frequency can be performed by expansion in the powers of the small parameter $k_B T_e/\varepsilon_b$. Accordingly, the squared frequency of the phonons in the excited solid linearly decreases as the electrons' temperature grows up as the first approximation:

$$\omega_{\text{ph}}^2 \approx \omega_{0,\text{ph}}^2 - k_B T_e / \varepsilon_{0,b} \tag{4.10}$$

Bi has been excited to the maximum electron temperature of 0.3 eV in the experiments (Garl et al., 2008). The experimentally measured frequency of A_{1g} mode constitutes 2.82 THz (compare with the frequency 3 THz at room temperature). Formula (4.10) predicts 2.78 THz indicating good semi-quantitative agreement.

4.1.2.3 Phonon's excitation imprinted into the transient optical properties

Swift but subtle laser excitation by the action of the electron pressure force generates standing waves of the coherent atomic vibrations, which were experimentally traced as the oscillations of the reflected probe. The dielectric permeability of any material depends on the polarization part and on the electronic part. The electronic part depends on the changes in the number of the conductivity electrons and on the electron–phonon collision rate. The electrons are heated to the maximum temperature at the end of the laser pulse, which is much shorter than the period of the phonons oscillations. Quick blow of the electronic pressure force excites the phonons oscillations. The phonon's amplitude explicitly enters into the electron–phonon collision rate thus affecting the electronic part of the time-dependent permittivity. Note that the number of the conductivity electrons and their velocity are practically constant during several picoseconds of gradually damped atomic vibrations. Thus, the sequence of the processes, finally ended by the reflectivity oscillations of the probe, is as follows. The fast laser blow slightly perturbs the amplitude of the phonons oscillations, $q(t) = q_0 + \Delta q(t)$, which results in the perturbation of the electron–phonon collision rate, $\Delta \nu_{\text{e-ph}}(t) \approx \nu_{0,\text{e-ph}} \frac{2\Delta q(t)}{q_0}$. Electron–phonon rate enters into the imaginary part of the permittivity and therefore into the reflection coefficient expressed through the Fresnel formula. The perturbation in the reflectivity induced by the laser-generated phonon vibrations reads

$$\Delta R(t) \approx \frac{1}{2}\left(\frac{\partial R}{\partial \nu_{\text{e-ph}}}\right)\Delta \nu_{\text{e-ph}}(t) \approx \left(\frac{\partial R}{\partial \nu_{\text{e-ph}}}\right)_0 \frac{\partial \nu_{0,\text{e-ph}}}{q_0}\Delta q(t) \tag{4.11}$$

It is clear from (4.11) why the reflectivity oscillates with the cold phonon frequency.

4.1.2.4 *Experiments*

The reflectivity of a probe beam with the time resolution of 100 fs and less was recorded in many experiments (Zeiger et al., 1992; Boschetto et al., 2008) at the absorbed energy density around and up to several times larger than the equilibrium enthalpy of melting. Previous discussion shows that transformation from the solid state to the liquid-like state under strong excitation occurs in a different way from the solid-to-liquid transition in equilibrium. The major difference is that under speedy excitation the optical properties defined by the electrons are changing first while the lattice responds only by the phonon's excitation. The atoms may change their space positions allegedly under the action of the electronic pressure force and in the course of the energy transfer from electrons to the lattice. However, the optical probe cannot trace changes in the atomic positions. Let us discuss the time-dependent measurement optical probe reflections from swiftly excited Bi and identify the similarities and differences between equilibrium and non-equilibrium material transformations.

Boschetto et al. (2008) measured the reflectivity of the probe beam from the single crystal of bismuth (111)-oriented in the direction to the pump laser. Pump laser was Ti: Sapphire (800 nm, 35 fs, 1 kHz repetition rate, energy up to 10 mJ per pulse), the small part of the main beam used as a probe. The reflectivity changes, ΔR, were observed with the accuracy $\Delta R/R \sim 10^{-5}$ and the temporal resolution of 35 fs (see Fig. 4.1).

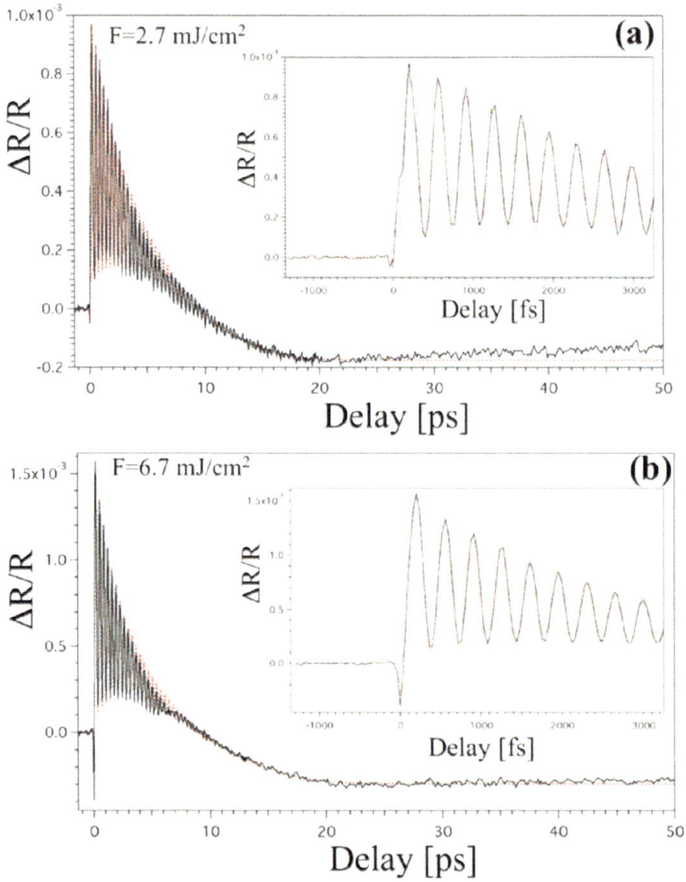

Figure 4.1 The measured (dark solid line) and the calculated (red dotted line) time-dependent reflectivity of the probe optical beam, $\Delta R/R$, at excitation fluence of 2.7 mJ/cm^2 (a) and 6.7 mJ/cm^2 (b); the zero point on the time scale indicates the arrival of the laser excitation pulse. The insets show the first 3000 fs after the pulse in details (from Boschetto et al., 2008).

First, the coherent atomic displacement is produced by the polarization force and electron pressure force that is recorded as the negative drop in the reflectivity. The reflectivity increases over the level for a cold solid and reaches its maximum during the time comparable to the cold phonon period (hundreds of femtoseconds) well after the end of the pulse. Then the reflectivity oscillates with the cold phonon frequency over the period depending on the pump energy with the gradually decreasing amplitude.

The phonon's amplitude steadily decreases while electrons transfer the energy to the lattice. Heating of the lattice and thermal expansion transformed initially harmonic vibrations of atoms into non-linear motion that manifests onset of the phase transformation. Observed dependence of the damping rate on the temperature is close to the probability per unit time of the optical phonon decay into two acoustic phonons. The inverse damping rate can be interpreted as a lifetime of phonon. The analysis of the reflectivity oscillations evidences that a solid experiences transition to a transient state with the optical parameters different from those for solid and liquid. This intermediate transient state is difficult to identify with the known equilibrium liquid state.

4.2 Fast Transformations in Non-Equilibrium: Transient Phase States—Neither Solid Nor Melt (Metals-Bi, Ga, Al; Dielectrics-Sapphire)

The major difference between phase transformations in equilibrium and under fast action of the intense laser pulse is the transient nature of the transformation in comparison to almost steady-state conditions of transformation in equilibrium. However, the common feature is that the deposited energy density is much less than the threshold for the any changes at the target surface (often referred to as damage). The material density during the equilibrium melting changes slightly. In majority of materials, it decreases in the melt. Only in five solids, the density in the liquid state slightly exceeds the solid density. During the melting, the optical properties are changing along with the loss of the long order (disordering). It is instructive to briefly review the main features of melting in equilibrium and then compare them to the succession of the processes during the fast transformation.

4.2.1 Road to Melting in Equilibrium: Sequence of the Catastrophes

The melting of the different crystals in equilibrium has features in common. It was established that successions of instabilities (catastrophes) are preceding the crystal disordering. The ultimate stability limit of a superheated crystal is attained when the entropy of a crystal equals to its liquid entropy. At this moment coined as the entropy catastrophe, a solid is disordered. It was found that the point defects concentration and the lattice heating are the major contributions to the catastrophic entropy value (Fecht and Johnson, 1988; Tallon, 1989; Fecht, 1992).

The point defects, vacancies, the inherent feature of equilibrium, are generated by the energetic particles from the high-energy tail of equilibrium distribution function (Fecht, 1992). If the vacancy contributions were neglected the critical entropy (and onset of disordering) would be achieved solely by superheating a crystal up to $3T_m$ (Fecht and Johnson, 1988). The superheating of high-quality defect-free crystals in equilibrium to temperature beyond its melting point without the solid melting is a well-known phenomenon.

4.2.1.1 Main features of the melting in equilibrium

The melting point in equilibrium, T_m, at a pressure, P_m, is the temperature at which the Gibbs free energies of the two phases are equal. At $T > T_m$, thermodynamic instability occurs, the transformation commences, and the cohesive (binding) energy decreases. Thus, the melting process leads to the changes in the inter-atomic potential. The standard enthalpy of fusion (specific heat of melting), ΔH_f, is the amount of thermal energy absorbed per unit volume necessary for the change of a material state from a solid to a liquid. Correspondent entropy of fusion is the increase in entropy due to the heating a solid to the melting point, $\Delta S_m = \Delta H_f / T_m$.

The major differences between solid and liquid are as follows: (1) The long-range order of the crystalline solid changes to the short-range order in a liquid; (2) the shear modulus in a liquid is zero; (3) the number of conductivity electrons in the conduction band may increase in a melt, as it happens in bismuth (see Comins, 1972).

The several melting criteria were introduced during the past century, which tried to establish the common features of transformation in all solids. These are the vibrational instability of Lindemann (1910), the elastic shear instability (rigidity catastrophe) of Born (1939); instability generated by the presence of defects (vacancies) of Gorecky (1977). Lennard-Jones and Devonshire (1939) were the first who treated the melting as transition from ordered to disordered state. The entropy catastrophe of Fecht and Johnson (1988) at $T > T_m$ sets the ultimate limit for the crystalline stability and allows connecting value of entropy to the critical temperature at which the catastrophic melting occurs.

Lindemann (1910) assumed (see also Pines, 1965) that any solid melts when the amplitude of phonons exceeds some threshold value being a small part of the inter-atomic distance. The Lindemann criterion does not establish a melting threshold. It indicates qualitatively on the onset of the vibrational instability.

Born (1939) defined the melting temperature as the state where the shear stress turns to be zero signifying the transition to the liquid state. Qualitatively the condition of zero shear modulus coincides with the onset of vibration instability of Lindemann. Note that both criteria are considering changes in atomic positions due to lattice heating.

The entropy catastrophe of Fecht and Johnson (1988) sets the ultimate limit for the crystalline stability at $T_{\mathrm{cr}} > T_m$. The critical temperature corresponds to the state where the entropy of a solid equals to that for its liquid. Beyond this point, a superheated solid cannot exist because entropy of solid at $T > T_{\mathrm{cr}}$ becomes larger than in liquid that is incompatible with the thermodynamic laws. Total change in entropy is a sum of contributions from electrons, lattice and thermal defects.

4.2.1.2 Critical entropy and critical temperature

Total change in entropy of the overheated metal is a sum of contributions from electrons, lattice and thermal defects. We shall measure entropy in units of the Boltzmann constant. The detailed analysis shows (Gamaly, 2011) that major contributions to the entropy of overheated metal in the non-equilibrium come from the isochoric heating of the lattice and from the thermal defects. Contribution from the electrons to the entropy changes in good metals is low. Indeed, the isochoric (constant volume) contribution to the entropy comes from the heat capacity of the degenerated electrons, $C_e = \pi^2 k_B^2 T_e / 2\varepsilon_F$:

$$\Delta S_e(T) = \int_{T_0}^{T} (C_e/T)dT = \pi^2 k_B^2 (T - T_0)/2\varepsilon_F$$

One can see that contribution of electrons heated to the melting temperature in good metals is small because the melting temperature is small compared to the Fermi energy. For example, the contribution of electrons to the entropy changes in Al at the melting point equals to $\Delta S_e \approx 0.023 k_B$ per electron.

A major contribution to the entropy comes from the lattice heating:

$$\Delta S_{\text{th}}(T) = \int_{T_0}^{T} (C_p/T)dT \approx C_p \ln(T/T_0).$$

Thus, the lattice heating in the good metals from the room temperature to the melting point contributes to the entropy change in a range, ΔS_{th} (T_m) = (3.5–5.5) k_B (see Appendix 3). We took the heat capacity approximately constant and equal to the Dulong–Petit value in the Boltzmann constant units.

Fecht (1992) pointed out that the destabilization of a crystal by heating is enhanced by the formation of non-equilibrium point defects, such as vacancies, anti-site defects and interstitials, which are intrinsic features of equilibrium characterizing static disorder. The number of defects corresponds to the number of broken bonds in a lattice and it is proportional to the number of atoms in high-energy tail of the Maxwell distribution, $n_{\text{vac}} \propto \exp(-\Delta H_v/k_B T)$; ΔH_v is the heat for the single vacancy formation. Increase of the defects concentration, c_d, from 10^{-3} (crystal at T_m) to 10% leads to solid-liquid transformation. This transition could be suppressed (or delayed) in the absence of defects.

The heat of formation for a vacancy is related to the equilibrium melting point T_m by the universal relationship,

ΔH_v = 9.28 × $k_B T_m$, established for a wide range of metals (Fecht, 1992). The total energy density of defects is estimated as follows:

$$E_{\text{def}} \approx \Delta H_v \cdot n_{\text{def}} \approx C_{\text{def}} \cdot \Delta H_v \exp(-\Delta H_v / k_B T)$$

Then the entropy change due to presence of the vacancies reads

$$\Delta S_{\text{def}} = \int_{T_m}^{T} dE_{\text{def}} / T \approx \int_{T_m}^{T} C_{\text{def}} \cdot \Delta H_v^2 \cdot \exp(-\Delta H_v / k_B T) dT / T^2.$$

Fecht (1992) calculated the change in the entropy arising from the presence of defects with the proportionality coefficient, C_{def}, found from the measurements, $\Delta S_{\text{def}} \approx 2 \times k_B$.

It is clear, then, that lattice heating and thermal defects provide key contributions to the critical entropy. The critical entropy for Al is ΔS_{crit} = 6.41 k_B. This value is reached at the temperature $T_m^{\text{vs}} \approx 1.38 \cdot T_m$ (Fecht and Johnson, 1988; Tallon, 1989). If the defect contribution is neglected, then the critical temperature corresponding to the entropy catastrophe in Al increases to $T_m^s \approx 3T_m$ (Fecht and Johnson, 1988). The critical entropy for a majority of metals is in a range $\Delta S_{\text{crit}} \sim$ 5–7 k_B. This value of the critical entropy establishes a mark for estimating the onset for non-destructive transformation in non-equilibrium conditions as well.

4.2.2 Ultra-Fast Material Transformations

The ultra-short pulse swiftly excites electrons to the average energy per electron several times larger than the melting point at the deposited energy density comparable to the enthalpy of fusion. The electron and lattice temperatures are equilibrated long after the end of the sub-picosecond pulse. Therefore, it is reasonable to ask ourselves: Could the energetic electrons initiate a solid-liquid-like transformation before the energy transfer to the lattice by the electron-to-phonon collisions? We would like to apply the criterion of entropy catastrophe used in equilibrium for defining the onset of phase transformation during very fast heating.

4.2.2.1 The legitimacy of the entropy concept for the description of rapidly excited metal

Two conditions should be fulfilled for making appropriate the application of the entropy concept for description of the fast material transformation. First, the distribution function that defines the thermodynamic parameters such as temperature, pressure, entropy, etc., should be set up. Second, temperature, entropy and time scale of the processes should satisfy conditions when the quantum fluctuations are negligible.

In a sub-system containing the only one sort of particles (electrons and lattice in a metal), the average energy in each sub-system establishes after few collisions. It is conventionally assumed that a single collision time is a characteristic time scale for establishing the main part of the distribution function and the average energy allowing the other thermodynamic parameters to be used for characterization of the processes. Time for establishing the distribution function in the electron sub-system is the inverse of the electron–electron collision rate, $t_{\mathrm{ee}} \approx \nu_{\mathrm{ee}}^{-1} \propto \omega_{\mathrm{pe}}^{-1} \varepsilon_F / \varepsilon_e$. Here, the plasma frequency is $\omega_{\mathrm{pe}} = (4\pi e^2 n_e / m_e^*)^{1/2}$, ε_F and $\varepsilon_e = k_B T_e$ are the Fermi energy, and energy of the electrons heated over the Fermi level. Example: In Al rapidly heated up to the enthalpy of melting ($\varepsilon_e = \varepsilon_F/50.56$), the electron distribution (and average temperature) establishes in ~ 2.6 fs. Thus, the thermodynamic description of the electrons entropy is legitimate few femtoseconds after the beginning of the pulse.

Likewise the time for the setting up the main part of the Boltzmann distribution (lattice temperature) is the inverse of the phonon–phonon collision rate, $t_{\mathrm{ph-ph}} \approx (\nu_{\mathrm{ph-ph}})^{-1} \cong \hbar / k_B T_L$. The lattice temperature in Al, where electrons were swiftly heated, remains low during the electron-to-lattice transfer time of a few tens of picoseconds thus keeping the distribution function close to its equilibrium shape. When the lattice is heated up to the melting point, T_m = 933.5 K, the time for the distribution function setting up, decreases from 18 to 8 fs. One can see that the time scales for establishing statistical distributions in other fast-excited metals are also in the range of a few femtoseconds.

The second condition for application of thermodynamics is that the entropy fluctuation during the process in question should exceed the quantum limit (Landau and Lifshitz, 1960):

$$\Delta S >> \hbar/(t \cdot k_B T)$$

One can see that quantum fluctuation of entropy at $T \sim 10^3$ K during the time $t \sim 100$ fs constitutes 0.07 k_B that is around one per cent of the critical entropy value. Therefore, one can use legitimately the entropy concept for the description of the ultra-fast transformations.

4.2.2.2 Entropy changes during the quick heating

The catastrophic disordering in equilibrium, melting, commences after three processes in succession, which increase the entropy to the critical value, are completed: the electrons' heating, the thermal point defects formation and the lattice heating. After the rapid heating, the electron temperature is several times larger than that in the lattice. The electrons' contribution into the entropy changes following strong and fast electron's excitation is three-fold. The first contribution comes from the isochoric electron temperature rise. It is equivalent to the electron's contribution into the equilibrium melting. The inhomogeneous electron's heating and the local modification of the inter-atomic potential are two other contributions to the entropy changes, which are specific for the swift and strong electronic excitation.

Therefore, it is worth estimating *the electrons' contribution into the entropy change* as a primary suspect for the non-thermal melting (before the temperature equilibration). The rapid electron temperature rise during the laser pulse results in an isochoric (constant volume) increase in the entropy of the degenerate electrons whilst the lattice remains cold, $\Delta S_{\mathrm{cr}}(T_e) = k_B^2 \pi^2 (T_e - T_0)/2\varepsilon_F$. One can see that the electronic contribution into the entropy increase even in the strong overheating, $T >> T_m$, is much less than the critical entropy of $\Delta S_{\mathrm{cr}} = (6\text{–}7) \times k_B$. For example, the strong overheating of electrons in Al to the temperature $k_B \times T_e = 9.4\, k_B \times T_m = 0.756$ eV results in an entropy increase of $0.32 \times k_B \approx \Delta S_{\mathrm{cr}}/20$.

The physical reason behind this effect is clear: Even at the strong non-destructive overheating, the electron energy is much less than the Fermi energy, $k_B \times T_e << \varepsilon_F$.

Swift electronic excitation, while atoms remain cold, may affect the atomic motion in two ways. First, the local excitation of electrons changes the inter-atomic potential increasing the inter-atomic spacing. Second, the electrons' heating in a skin-layer is essentially inhomogeneous imposing the electronic pressure gradient force on atoms and therefore shifting atoms from the equilibrium positions. Both contributions even in strong overheating ($T_e \approx 9 \times T_m$) give the entropy changes less than 0.1 of the catastrophic entropy value. *Strong and fast electron excitation alone, therefore, cannot produce the entropy changes comparable with the catastrophe value that signifies a solid disorder (or transformation into some transient state).*

Lattice contribution to the entropy change is a maximum at the electron-to-lattice temperature equilibration time. This time decreases when the deposited energy density grows up. One can easily see that in spite of the fact that maximum electron temperature in swiftly heated metal is always several times larger than the maximum lattice temperature, the lattice contribution into the entropy change is overwhelmingly larger. Therefore, *the fastest time for the catastrophic lattice disordering is the electron-to-lattice energy transfer time and the lattice overheating to $T_L = 3 \times T_{\text{melt}}$.* Example: Aluminium ($\varepsilon_F$ = 11.63 eV; $n_e = 1.806 \times 10^{23}$ cm^{-3}; T_m = 933.47 K; $k_B \times T_m$ = 0.08 eV). At the deposited energy density equal to the enthalpy of melting, ΔH_f = 1.07 kJ/cm^3 the maximum electron temperature in Al reaches $T_{e,\max} = 2.9 \times T_{\text{melt}}$ keeping the lattice temperature at the melting point after the temperature equilibration. The electron–phonon momentum exchange rate, $\nu_{\text{e-ph}} = 1.5 \times 10^{15}$ s^{-1}, leads to the energy exchange rate, $\nu_{\text{energy}} = 3 \times 10^{10}$ s^{-1}, and temperature equilibration time of 33.3 ps. Three times overheating to $E_{\text{abs}} = 3 \times \Delta H_f$ results in the lattice temperature of $T_{L,\max} = 3 \times T_{\text{melt}}$, which produces the entropy change equal to the catastrophe value. This state is reached at the temperature equilibration moment 19 ps after the end of the pulse.

Generation of the thermal point defects is another major contribution into the entropy increase leading to disordering. However, point defects formation in metal needs high-energy atoms in the Boltzmann distribution, $\Delta H_{\text{vac}} = 9.28 \cdot k_B \cdot T_m$. Therefore, the point effects formation time is much longer than the electron–lattice energy transfer time.

The analysis above based on the thermodynamic laws demonstrates that the *sole electronic excitation in a metal cannot induce the material disordering*. In other words, *non-thermal melting*, e.g., disordering of the cold lattice before the energy transfer from the hot electrons, is impossible due to the violation of the thermodynamics laws. However, the electronic excitation should lead to the changes in the transient optical properties of laser-excited solid, which were traced experimentally with the ultra-fast optical probes. Atomic displacement and possible loss of the long order might be observed with the ultra-fast x-ray or electronic probe.

4.2.3 Novel Transient Phase States—Neither Solids Nor Melts

Now we discuss the experiments, in which intense sub-picosecond laser excited a solid at the absorbed energy density below the damage threshold. The relaxation history of the excited material return to the ambient conditions has been observed with single or double probes restoring the time-dependent material properties with high time resolution. The analysis presented above demonstrated that fast non-destroying excitation might produce subtle changes in the material structure. On the other hand, excitation of the electron sub-system strongly affects also the rate of the electron–phonon interaction and therefore the transient optical properties of the excited material. Therefore, in what follows, we concentrate on the probe-detected transient optical properties.

The energy density for producing similar transformation in metals and dielectrics in equilibrium are of the same order of magnitude. For example, the enthalpy of fusion for ice equals to 0.335 and 0.5 kJ/cm^3 for Bi, at the low end up to 0.975 kJ/cm^3 for InSb and 1.86 kJ/cm^3 for Copper at the high enthalpy limit. One needs the order of magnitude higher laser fluence (J/cm^2) for depositing in dielectric the absorbed energy density comparable to that in a metal.

4.2.3.1 Excitation of metals/semi-metals—Ga, Bi, Al

Gallium

The reversible phase transformation of α-gallium excited by the femtosecond laser pulses (150 fs, 775 nm) has been studied by measuring the time-dependent reflectivity of two identical simultaneous femtosecond probes set at two different angles (Uteza et al., 2004). The absorbed energy density was in a range (0.5–2) × H_f, where H_f = 0.479 kJ/cm^3 is the enthalpy of fusion α-Ga in equilibrium. The real and imaginary parts of the permittivity of solid α-Ga and liquid were measured in equilibrium. It was found that permittivity is described well by the Drude-like formula. Two independent simultaneous measurements at any time moment allow recovering the real and imaginary parts of the transient permittivity with ~ 200 fs time resolution. Two parts of the permittivity in turn depend on two parameters: the ratio of the squared plasma frequency to the laser frequency $(\omega_{pe}/\omega)^2 = n_e/n_{cr}$ being the ratio of the electron density to the critical density, and on the ratio of the electron–phonon effective collision rate to the laser frequency, ν/ω. Thus, the time history of the collision rate and the number density of the conductivity electrons during 25 ps after irradiation have been recovered (Fig. 4.2). It was found first that optical properties of swiftly excited α-Ga begin changing at the deposited energy density half of the equilibrium enthalpy of fusion. The real and imaginary parts of the permittivity were still not approaching the liquid values even after two times increase in the deposited energy density over the enthalpy of fusion,. Thus, intense femtosecond laser transforms α-Ga at the deposited energy density twice bigger the enthalpy of fusion into a phase state whose optical properties are different from those for the crystal and for the liquid gallium.

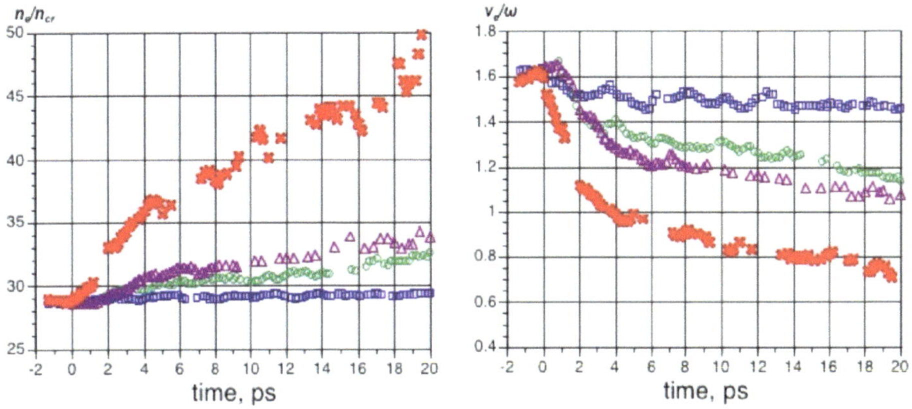

Figure 4.2 Time history of the electron density $n_e/n_{\text{cr}} = (\omega_p/\omega)^2$ and electron–phonon collision rate ν_{opt}/ω at different pump fluences: 1.4 mJ/cm^2 (squares); 1.8 mJ/cm^2 (circles); 2.2 mJ/cm^2 (triangles) and 2.85 mJ/cm^2 (crosses) laser fluences (From Uteza et al., 2004).

Uteza et al. (2004) explained such a state as a mixture of liquid and solid phases. The alternative interpretation suggests the formation of a homogeneous transient phase with a structure different from that in solid and liquid Ga. The single-probe measurements indicate that the excited gallium returns to the initial state in nanosecond time.

Transient state of fs-laser excited bismuth

Time history of bismuth crystal transformation excited by the pump laser (50 fs, 775 nm, 6.9 mJ/cm^2) was observed measuring the time-dependent reflection of two synchronized optical probes set to reflect at two different angles, 19.5° and 34.5°. The absorbed energy density delivered by the 6.9 mJ/cm^2 pump equals 1.24 kJ/cm^3, which is 2.5 times larger than the equilibrium enthalpy of Bi melting. Therefore, Bi was overheated to the maximum lattice temperature around 1,360 K (melting point, $k_B T_m$ = 544.7 K). The reflectivity changes were measured with the accuracy of 10^{-5} (Rode et al., 2009; Boschetto et al., 2010). The pump beam was *p*-polarized, whereas the two probes were *s*-polarized. The numerical solution to two equations for two reflection coefficients expressed by the Fresnel formulae allows recovering the time dependence of the real and imaginary parts of the permittivity (Fig. 4.3).

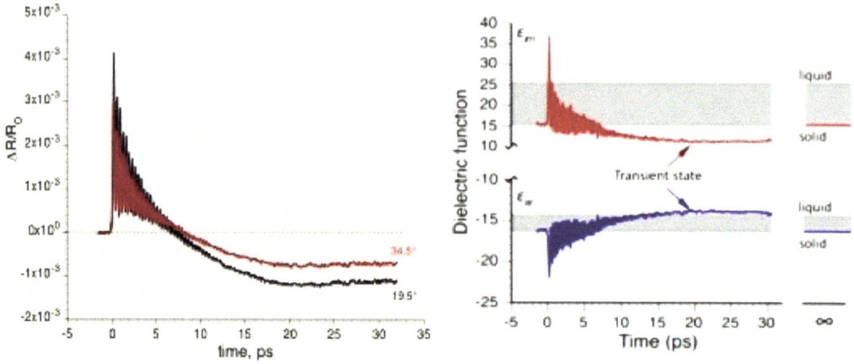

Figure 4.3 Time-dependent reflectivity of the two simultaneous probes (left plot) and recovered real and imaginary parts of the dielectric function of Bi (right plot) excited at the absorbed energy density almost three times of the melting enthalpy and recorded with 60 fs resolution (after Rode et al., 2009).

The reflectivity oscillations with the A_{1g} optical phonon's frequency were damped down in ~ 10 ps. One can see that the maximum of the real part of the permittivity was achieved at the end of the pulse. The imaginary part of the permittivity is a maximum after the end of the pulse decreasing to almost constant value in the intermediate state, which lasts from 15 to 30 ps. The permittivity in the transient state reads $\varepsilon_{\text{trans}} = -1.38 + i1.13$, drastically different from that in solid $\varepsilon_{\text{solid}} = -16.25 + i15.4$ and liquid, $\varepsilon_{\text{liquid}} = -14.35 + i25.5$. Recovery of the electron number density (n_e/n_{cr}) and collision rate (ν/ω) from the Drude-like permittivity well describing the solid and liquid Bi in equilibrium gives the unexpected results: Electrons' density appears to be 1.5 times lower than that in solid and almost 2 times lower than that in liquid. The collision rate also drops lower than in the equilibrium solid and liquid. The analysis of these experiments (Gamaly and Rode, 2013b) shows that the cooling time based on the diffusivity coefficient measured in equilibrium and non-equilibrium conditions is around 3.6 ps. However, the permittivity of the excited Bi drops faster than it might be due to the simple cooling. The consistent interpretation of these experiments is still absent. The amazing transient optical properties might be an indication on the unusual structural and electronic changes. The excited Bi returns to the initial conditions in ~4 ns (Boschetto et al., 2010).

One might suggest that during the ultra-fast cooling the excited crystal arrives in the short-lived energy minimum corresponding to a transient crystalline phase state with the optical properties distinct from those as in the initial solid or in disordered liquid. Analysis from Gamaly and Rode (2013a,b) indicates that the entropy of excited bismuth is lower than the catastrophic value threshold for disordering because the fast cooling prevents the formation of the thermal point defects crucial for the lattice randomization. It should be noted that an interpretation of the transient state in Bi as a mixture of the solid and liquid states is impossible. Summing up, one might conclude that the discussed experiments are the strong indication of the existence of unusual transient state with the peculiar optical properties in the rapidly laser-excited metals.

4.3 Mixed Dielectric/Metal Properties in Swiftly Excited Transparent Dielectric (Sapphire Die-Met)

Wide-band gap dielectrics are transparent for the low-intensity light in the optical wavelength range (up to wavelength, 100–150 nm). Absorption increases when the high-intensity laser beam transfers electrons from the valence to conduction band. Hence, it is necessary to deposit into a dielectric an order of magnitude larger energy density than in metals in order to achieve ionization and absorption first and then transient melting-like transformation.

The real part of the permittivity of the excited dielectric decreases due to ionization remaining positive while the imaginary part responsible for absorption grows up. At this level of the absorbed energy density, the transient states of excited dielectric exhibit simultaneously both dielectric and metallic properties. These states are termed the "Die-Met" states (Gamaly and Rode,

2018). The excited material is transparent and conductive at the same time. At some threshold fluence and absorbed energy density (see Chapter 3), the real part of the permittivity turns to zero and becomes negative at over the threshold fluence.

Swift excitation of transparent dielectrics by ultra-short intense laser pulse leads to ultra-fast re-structuring of the electronic landscape and generates variety of transient material states, which are continuously reshaped according to the changing pulse intensity.

It is interesting to investigate the transient properties of the excited material in the unexplored range of the permittivity, $0 \le \varepsilon_{\mathrm{re}} \le 1$. The number density of the conduction electrons at $\varepsilon_{\mathrm{re}} = 1$ is estimated as $n_e \approx (\varepsilon_0 - 1) \cdot n_{\mathrm{cr}}$.

Let us consider, for example, sapphire (Al_2O_3, $\varepsilon_0 = 3.13$) excited by an 800 nm laser ($n_{\mathrm{cr}} = 1.75 \times 10^{21}$ cm^{-3}). The number density of the conduction electrons in the state, where $\varepsilon_{\mathrm{re}} = 1$, equals $n_e \approx 2.90 \times 10^{21}$ cm^{-3}. That is much higher than that in pure semiconductors (10^{13}–10^{16} cm^{-3}), and in semi-metals (the highest concentration is in arsenic, $n_e = 2 \times 10^{20}$ cm^{-3}) (Kittel, 1996).

The electron–ion collisions ($\nu_{\mathrm{e-i}} = 7.6 \times 10^{14}$ s^{-1}) dominate in the permittivity range $0 \le \varepsilon_{\mathrm{re}} \le 1$. Conductivity, σ, of free electrons in the state $\varepsilon_{\mathrm{re}} = 1$ reads

$$\sigma = \frac{(\varepsilon_0 - 1)}{4\pi} \cdot \frac{\nu_{\mathrm{e-i}}}{(1 + \nu_{\mathrm{e-i}}^2/\omega^2)} \approx 10^{14}\ \mathrm{s}^{-1}.$$

This value of the conductivity is much lower than in metals ($\sigma \sim 10^{17}$–10^{16} s^{-1}) (Kittel, 1996) but larger than in the semi-metals. Thus, the laser-modified sapphire has the mixed metal/dielectric properties in the absorption layer with the thickness of, $l_s = 394$ nm.

Note that in the permittivity range $0 \le \varepsilon_{\mathrm{re}} \le 1$, the optical properties depend mainly on the electron density. Therefore, the temporal and spatial mapping of transient permittivity in the laser-excited dielectric reproduces the pattern of intensity distribution in the interaction region. In what follows, we discuss the transient states of matter generated in the first stage of the interaction when the real part of the permittivity remains positive.

Along with the changes in the permittivity and conductivity, the modification of the electronic landscape creates excited states in the forbidden band gap. Allegedly, at the ionization threshold, the band gap is forced to close. Ionization of a crystal starts with the threshold-free multi-photon process. This process first creates the excited states of different lifetime within the band gap. Afterwards, the electrons are transferred from these intermediate states to the conduction band.

The gap between valence and conduction bands decreases during the electron transfer. The dependence of the indirect band gap, for example, in silicon, on the conduction electron density supports the above assumption: The band gap decreases by almost 0.2 eV (from 1.1 to 0.9 eV) due to heating to 1000 K, while the electron number density increases up to 2×10^{18} cm^{-3} (Lautenschlager et al., 1985). The decrease in the band gap should affect both the multi-photon ionization rate and the impact ionization as well. Keldysh accounted for the ionization from the excited states in the calculations of MPI ionization probability (Keldysh, 1965).

Allegedly, the lifetime of the excited states might be limited by the time period necessary for the energy transfer from electrons to ions. For the laser-excited sapphire, the energy transfer time is around 30–40 ps, which is in qualitative agreement with the experimental observation (Winkler et al., 2017).

Summing up, the transient material states possess the permittivity different from that in the parent solid or liquid, a shrinking band gap and excited electronic levels within the gap, along with a modified inter-atomic potential.

4.4 Inverse Population in Laser-Excited Sapphire and Silica (Winkler et al., 2017)

Recently, the amplification of the probe beam transmitted through the intense laser excited sapphire has been observed. Apparently, this is the experimental evidence of the existence of the excited electron states inside the band gap of the swiftly heated sapphire.

The intense short pump laser (2.6 J/cm^2; 30 fs; 800 nm; ~10^{14} W/cm^2) converts the sapphire layer into transient state with time/space varying permittivity and shrinking band gap. The excitation is strong: The deposited energy density is close to the ablation threshold. Hence, the permittivity of the excited sapphire varies from the positive values at the outer boundary of the focal spot to the negative real part values of the metal-like state in the focal spot centre. The band gap shrinks and eventually collapses after conversion to plasma in the centre of the focal spot.

The state of the pump-excited sapphire has been followed by the measurements of 400 nm probe transmission (~1/30 of the pump pulse energy) through the sapphire layer. It was found that 300 fs after the pump's excitation the probe energy increases in 1.5 times. The effect of amplification occurs only when the pump fluence is higher than the threshold for strong-field excitation of about 0.8 J/cm^2. At such fluence, the number density of the conductivity electrons significantly changes permittivity and modifies the band gap creating the excited energy levels.

The state of the excited sapphire creating the probe transmission's amplification is sustained for 50ps. Probably this period indicates the lifetime of the excited states in the shrinking but not yet collapsing band gap.

The excited sapphire consists of two parts: In the centre of excited spot, the absorbed fluence is a maximum, sapphire is converted to plasma becoming opaque for the probe. In the peripheral ring of the focal spot, where the absorbed fluence is well below the plasma threshold, the real part of the permittivity is positive and the excited sapphire is transparent for the probe. It is highly likely that the probe amplification occurs in this region. Note that in this ring, the band gap width decreases along with the decreasing real part of the permittivity in the direction to the centre of the beam. It is reasonable to suggest that the excited states are created inside the initially forbidden gap during the band gap reduction. Therefore, the interaction of the probe beam with this outer part of the focal spot of excited dielectric may result in lasing (or it might be even a lasing without inversion). Long lifetime of this state, capable for the amplification of the probe, suggests the long relaxation time for electrons moved into conduction band and keeping the band gap shrunken but not fully collapsed in the low excited part of the focal spot. It is, however, likely that sapphire in the centre of the focus was transformed into the disordered plasma state.

There is not yet the full understanding and interpretation of the probe amplification phenomenon. We present a couple of existing interpretations. Winkler et al. (2017) suggest that "three-photon stimulated emission is responsible for the emission and amplification." They also claim that three-photon absorption (4.65 eV, three pump's photons of 1.55 eV) is responsible for the ionization of sapphire and glasses. However, it is well established that multi-photon ionization (MPI) and ionization by the electron impact in sapphire and glasses are responsible for the process of dielectric's transformation. To ionize sapphire (band gap of 9.25 eV) by the MPI process, one needs approximately 5–6 photons with energy of 1.55 eV (800 nm). Therefore, this interpretation looks unconvincing.

The observed transmission amplification might be interpreted as two photons' gain due to excitation by the pump laser (1ω) and then simultaneous emission of two photons (amplification) triggered by the probe laser (2ω). As indicated by Reichert et al. (2016), three-photon absorption is a process preventing the conversion of 2P absorption into a 2P gain. However, this interpretation cannot explain the delayed amplification.

The amplification of the probe beam was found in the similar experiments with fused silica. It was found (Winkler et al., 2017) that the capability for amplification disappears rapidly after the end of the excitation. The researchers attributed this dramatic difference with sapphire to the rapid decrease in the number of the conductivity electrons.

The results of similar coherent probe amplifications in sapphire (aluminium oxide, band gap 9.25 eV) and silicon oxide (silica, band gap 7.6 eV) suggest a general nature of the phenomenon common for the wide-band gap dielectrics. More theoretical and experimental studies are needed to fully understand this interesting phenomenon. It is worth noting that long-lived polarization coupling between the ground state and excited states of nitrogen excited by short intense laser pulse in a gaseous phase was experimentally observed (Mysyrowicz et al., 2019). The phenomenon was interpreted and quantitatively explained as lasing without population inversion in good agreement with the experimental data.

The complex 3D structure of the permittivity makes a transparent part of the excited dielectric (at $\varepsilon_0 > \varepsilon_{re} > 0$) optically active. The probe electromagnetic beam transmitted through the excited dielectric gets a twisted trajectory and accrues the geometric phase while passing through such a medium. Both the phase of the beam and the rotation of the polarization plane depend on the 3D permittivity structure. Measuring the transmission, polarization and the phase of the probe beam allows one to identify quantitatively these new transient states.

4.5 Summary

- The non-destructive interaction (intensity $\sim 10^9$ W/cm^2, fluence $\sim$ mJ/cm^2) allows one to gently modify the optical properties of laser-excited bismuth through the excitation of the coherent phonons. The subtle reflectivity oscillations of the probe beam with the cold phonons frequency were measured demonstrating the ability for precise control over the excited material properties.
- Swift non-destructive excitation at the energy density around kJ/cm^3 (enthalpy of melting) transforms metals/semi-metals/dielectrics to short-lived unconventional state with the optical properties different from either solid or liquid in equilibrium. Measured lifetime of these states is from hundreds of picoseconds to nanosecond.
- The intensity of short 400 nm probe beam transmitted through the dielectrics (Al_2O_3 and SiO_2) excited by the intense, 800 nm, pump has been significantly amplified. This is the evidence that a dielectric excited below the ionization threshold converts into the unconventional transient state with the excited levels created allegedly in the modified band gap.

Chapter 5

Ablation of Metals and Dielectrics

The atomic bonds are broken and a solid loses its integrity if the energy in excess of the cohesion energy per atom is delivered into a solid. The energy density necessary for evaporation in equilibrium, enthalpy of evaporation, H_{vap}, is in a range of a few tens of kJ/cm^3. At such energy density, any atom gets sufficient energy for breaking the inter-atomic bonds. In equilibrium, it occurs during the time much longer any relaxation times establishing equilibrium. The unbound atoms can evaporate from a free surface.

In order to produce evaporation by the short laser pulse in the non-equilibrium conditions the larger energy density should be delivered to a solid. The laser energy is absorbed by electrons and then transferred to atoms (ions) forcing the bonds breaking. The phenomenon of removing the unbounded ions by the short laser action from the surface of any material was coined by laser ablation in order to emphasize its differences from evaporation in equilibrium conditions.

The processes of the laser ablation of metals and dielectrics are different. The dielectrics should be ionized first ensuring the absorption of the incident radiation. Then different processes of energy transfer from electrons to ions take place making the

Femtosecond Laser-Matter Interactions: Solid-Plasma-Solid Transformations at the Extreme Energy Density
Eugene Gamaly

ISBN 978-981-4877-40-4 (Hardcover), 978-1-003-25661-8 (eBook)
www.jennystanford.com

features of the ablation processes dependent on the material and the laser parameters. It is possible to deliver the energy density well in excess that necessary for breaking structural bonds inside the transparent crystal. In this case, the succession of the processes during and after the interaction is different from the interaction with the surface. We consider these phenomena in the next chapter.

This chapter is devoted to a thorough description of the material removal in the non-equilibrium conditions. We discuss the important features of the phenomenon: the ablation mechanisms, and thresholds, ablation rates, ablated plume composition. The chapter is organized as follows. First, the evaporation in thermal equilibrium is revisited in order to remind a reader about the main features of the process in equilibrium. Then, the mechanisms of the non-thermal ablation, thresholds for ablation to commence in metals and dielectrics, ablated rates, mass and depth of ablated material as function of laser and material parameters are discussed in succession. The problems of the control over the ablated plume content in a single-pulse ablation are followed by a discussion of the interaction between multiple successive pulses hitting the same spot in high-repetition-rate regime. The effect of ambient gas on the ablation threshold and rates is then discussed.

5.1 Evaporation in Equilibrium

Let us revisit the evaporation in thermal equilibrium and clarify the differences with the non-equilibrium ablation by short-pulse laser.

A solid at room temperature is in a state of mechanical equilibrium: The inter-atomic forces of attraction and repulsion are mutually balanced. Minimum of the inter-atomic potential in equilibrium equals the binding energy, $U_{\min}(d) = -\varepsilon_b$, where d is the inter-atomic distance in equilibrium. The energy of evaporation per atom (the enthalpy of vaporization per atom) equals the cohesion energy. However, the boiling temperature (in the energy units) is more than 10 times lower the binding energy (Table 5.1). This is a direct indication that a solid heated to the boiling temperature ejects atoms from the high-energy tail of the equilibrium distribution function. Of course, in real evaporation the role of defects, inhomogeneity and impurities is essential. Here, we are considering the equilibrium evaporation of the pure ideal solid.

Table 5.1 The enthalpy of equilibrium vaporization per atom for metals

	Al	Cu	Fe	Pb
$\Delta H_{\text{vap}}/n_a$, eV	3.05	3.11	3.53	0.564
ε_b, eV	3.065	3.173	3.695	1.795
$k_B T_{\text{boil}}/\varepsilon_b$	0.071	0.07	0.0631	

One can see that for good metals the ratio of boiling energy per atom to the binding energy is less than 10%. That is close to the content of the high-energy particles in the distribution tail. Equilibrium assumes that evaporation proceeds during a period of time much longer than all the relaxation times. Thermodynamic equilibrium during the solid-to-vapour phase transition also means that temperature, pressure and chemical potential at the solid–vapour interface are the same for both phases.

The equilibrium evaporation rate of a solid into vacuum, i.e. the number of atoms leaving the unit surface per unit time, at the temperature T_{vap} follows from the conventional thermodynamics (Landau and Lifshitz, 1980):

$$(n_i v_i)_{\text{equilbrium}} \propto n_{\text{vap}} \cdot \left(\frac{k_B T}{M}\right)^{1/2} \cdot \exp\left(-\frac{k_B}{\varepsilon_b} T\right), \tag{5.1}$$

where n_{vap} is the characteristic density of the saturated vapours near the solid–vapour interface (that can be obtained only from experiment). The equilibrium evaporation rate has a clear physical meaning. The pre-exponential factor presents a free flow of evaporated atoms through the solid–vapour interface. Thus, only a small amount of particles with the energy $\sim\varepsilon_b$ from the high-energy tail in the Maxwell distribution evaporates from a solid in the conditions when solid and vapour are in equilibrium

(Landau and Lifshitz, 1980). The rate of evaporation in equilibrium is low because $k_B T_{\text{boil}} << \varepsilon_b$. The equilibration goes much faster than evaporation. In the non-equilibrium conditions, the time for establishing full equilibrium distribution including the high-energy tail is the longest of all relaxation times.

5.2 Major Relaxation Processes under Swift Short Pulse Excitation

(1) It was demonstrated experimentally that in order to ablate atom by the laser action, the energy density around 10^5 J/cm^3, well in excess of the equilibrium enthalpy of vaporization, should be delivered into a solid. When the intense laser is switched on, an electron is dragged by the electric field through the potential barrier, being stripped off from any atom. Freed electrons oscillate in the HF applied field and the oscillation energy is converted into the electron's thermal energy via collisions. All absorbed energy is confined in the thermal energy of free electrons during the ultra-short pulse while ions remain at the same temperature and spatial positions as in the initial equilibrium state. During the intense laser/matter interaction, the electron's energy has three transient components: the energy of oscillations in HF field, the thermal energy, and the Coulomb energy of the electrical charges attraction/repulsion. The interaction of charges after the ionization turns into the plasma mode. In any quasi-neutral (n_e = Zn_i) plasma all charged particles are oscillating due to density fluctuations in the electrostatic field of the Coulomb interaction between the charges (see Chapter 1). Plasma created is non-stationary and non-equilibrium, while the thermal energy of electrons is comparable to the Coulomb energy. Therefore, the Coulomb potential is unscreened. We assume that plasma is simple (fully ionized and single charged), i.e. the number of electrons equals that of ions, $n_i = n_e = n$. Then from the equation, describing the electrons oscillations, $\frac{dv_e}{dt} = \omega_{\text{pe}}^2 \times r_e$ follows that the characteristic time

for establishing the oscillation of the newly created electrons is the inverse electron plasma frequency, $t_{C,e} \approx \omega_{\mathrm{pe}}^{-1}$ ($\sim 10^{-16}$ s). Similarly, the ions start oscillating for a much longer time, $t_{C,i} \approx \left(\frac{M}{m_e}\right)^{1/2} \times \omega_{\mathrm{pe}}^{-1}$. This time equals 22–58 fs for ions with masses in a range M = 27–183.84 a.u. The ions hold the initial positions during much longer time, $t_{C,I} >> t_{c,e}$.

The Coulomb coupling between the electrons and ions pushing the ions oscillate appears to be the fastest process of electron/ions interaction after the conversion a solid to plasma. Note that the electron–ion momentum transfer collisions responsible for the absorption of the laser energy by electrons are a few tenths of femtosecond short. Therefore, even during the intense pulse shorter $t_{C,I}$, the electron's thermal energy may increase over the work function allowing the electrons to escape a solid. We consider later how it is possible to remove atoms from a solid by the extremely short pulses $t_p < t_{C,I}$.

(2) During the pulses longer than the time of electrons/ions Coulomb coupling but shorter than the electron-to-ion energy transfer time, $t_{C,I} < t_p < t_{\mathrm{energy}}$, electrons are attached to ions by the force of attraction. Therefore, electrons, gaining the energy in excess of the sum of the work function and the binding energy, escape the solid dragging the ions off the surface by the mechanism of so-called electrostatic ablation, sometimes dubbed as "cold" ablation. The last definition emphasizes the fact that the ions have no time to be heated.

This process is the non-equilibrium variety of the thermal emission in equilibrium. Indeed, in equilibrium ($T_e = T_i$), the number of heated electrons leaving the surface exponentially depends on the work function, w_{esc} (threshold energy for the electron emission from the metal surface) (Kittel, 1996):

$$n_e = 2(m_e k_B T_e / 2\pi\hbar^2)^{3/2} \exp\{-w/k_B T_e\}$$

The work function for metals is in a range 3–5 eV. Thus, in the considered non-equilibrium state condition, $k_BT_e \approx w_{esc}$, signifies the threshold for the electron emission from the laser-excited surface.

(3) The electrons are reaching the maximum temperature to the end of the pulse. The electron–ion momentum exchange rate at its maximum is estimated as, $\nu_{ei} \approx v_e/r_a$. Here, $r_a = (3/4\pi n_a)$ is the average inter-atomic separation. For $n_a = 10^{23}$ cm^{-3}, $v_e \approx 10^8$ cm/s, one gets $\nu_{ei} = 7.48 \times 10^{15}$ s^{-1}. The electron-to-ion energy transfer time via the Coulomb collisions is

$$t_{ei}^{energy} = \frac{M}{m_e \times \nu_{ei}}.$$

For silicon ($M \approx 28$ a.u.), the energy transfer time, $t_{energy} \approx 6.8$ ps, is much longer than the Coulomb coupling. The process of the energy transfer from electrons to ions via the collisions leads to the electron-to-ion temperature equilibration and establishing the main part of the Maxwell distribution for ions. However, the ion distribution function remains different from the full Maxwell form due to the absence of the high-energy tail. The high-energy tail needs the longest time to be built.

Another process starting simultaneously with the electrons' heating is the electronic heat conduction. Absorption length in plasma created at the deposited energy density close to the ablation threshold is around a few tens of nanometre, electron's velocity $v_e \approx 10^8$ cm/s. Hence, the diffusion coefficient $D = v_e^2/3\nu_{ei} \approx 0.4$–$0.5$ cm^2/s. The characteristic cooling time for the area comparable to the absorption length is $t_{cool} \approx l^2/D \approx 40$–$50$ ps.

After the electron–ion temperature equilibration, the heat conduction process during the long (nanoseconds) pulse action dominates. Hence, the threshold for material ablation by the nanosecond-long pulses becomes the laser-pulse-duration dependent.

It appears that the process of ion removal from the laser-excited surface is different for three different pulse durations:

(1) extremely short pulse, $t_p < t_{c,i}$ (less than a couple of tens of a femtosecond) (2) ultra-short pulse shorter than the time for the energy transfer from electrons to ions by collisions, $t_{C,i} < t_p < t_{\text{ei}}^{\text{energy}}$ (a few tens of a femtosecond up to a hundred ps) and (3) long pulse where the electron heat conduction dominates the process (larger than a few tenths of a nanosecond). Below we estimate the ablation thresholds for three types of the pulse duration.

5.3 Short Pulse Ablation Thresholds

In the laser-surface interaction experiments, the focal spot diameter (microns) is much larger than the absorption depth (a few tens of a nanometre) making one-dimensional description a good approximation. Then the minimum amount of removed material from evenly irradiated surface is the monoatomic layer.

Definition of the ablation threshold: The ablation threshold is then defined as a minimum amount of energy necessary to deliver at the unit surface (fluence) for the removal of one mono-atomic layer. In this definition, the "damage threshold" and "ablation threshold" notions are identical and objective.

5.3.1 Ablation of Metals

5.3.1.1 Extremely short pulses of duration shorter $t_{c,i}$

Let us consider the action on a metal of the intense pulse shorter than the time necessary for the Coulomb force to affect an ion, $t_p < t_{C,i}$. It is well established experimentally and understood theoretically that for the pulse duration from a few femtoseconds up to hundreds of femtoseconds, the ablation threshold for metals is independent on the pulse duration. This means that the average intensity per pulse is growing up with the pulse shortening, $I \sim (t_p)^{-1}$. For metals, the ablation threshold fluence by the femtosecond pulses is in a range 0.3–0.5 J/cm^2. Thus, for 10 fs pulse, the intensity is in a range (0.3–0.5) × 10^{14} W/cm^2.

The laser field swiftly pulls the electrons off through the potential barrier (work function, w_e) apart of the surface. The charge balance inside a material is destroyed and the force of the Coulomb repulsion ejects ions from the metal (see Fig. 5.1).

The process of the electron's ejection from a solid is the tunnelling of electron through the barrier with the height equal to the work function, w_e. The tunnelling by HF electric field is effective if the oscillation energy of electron in the field exceeds the height of the energetic barrier $4\varepsilon_{osc} > w_{esc}$. This condition also indicates that the time for passing the barrier by electron, t_{bar}, is smaller than the field period, $t_{bar} \times \omega > 1$. The width of the energetic barrier decreases as the amplitude of the applied laser field increases, $l_{bar} = w_e/eE$. The work function of metals, w_{esc}, is in a range of 3–5 eV. The width of the 4 eV wide barrier in the laser field, created at the intensity 10^{14} W/cm^2, equals $l_{bar} \approx 1.46 \times 10^{-8}$ cm. The electron's oscillation velocity in the laser field with intensity 10^{14} W/cm^2 is $v_e \approx 10^8$ cm/s (laser wavelength ~ μm). Thus, the time for the pass through the barrier amounts to $t_{bar} = l_{bar}/v_e \approx 10^{-16}$ s $<< \omega^{-1}$. The process of the ablation consists of two stages: (1) electrons escape from the surface, and (2) the Coulomb explosion ejects ions.

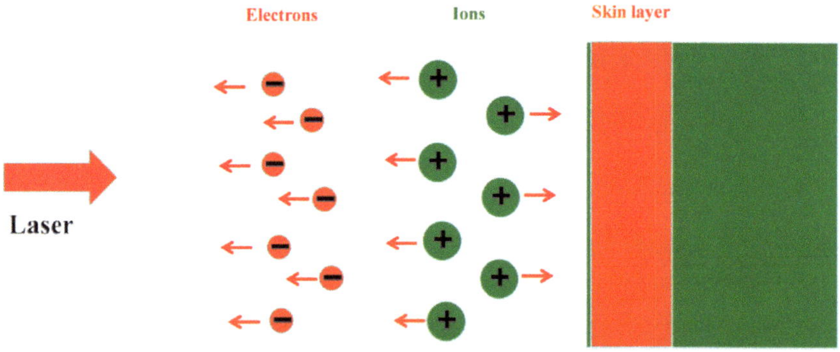

Figure 5.1 Energetic electrons, gained the energy during a laser pulse shorter than the inverse ion frequency time, then move off the surface to the distance where the Coulomb attraction to ions is ineffective. Bare ions repulsion removes some ions from the surface.

Let us look, for example, at the dynamics of the removal of Al ion by the action of the extremely short intense pulse, $t_p < t_{Ci}$ Taking $n_a^{Al} = 6.02 \times 10^{22}$ cm^{-3}, one gets $r_0 = 1.58 \times 10^{-8}$ cm and $t_{Ci} = 9.28$ fs, assuming that one electron per ion is ejected (ion is singly charged). The Coulomb energy of the ion–ion repulsion is $\varepsilon_{Coul} = e^2/r_0 = 9.1$ eV.

The corresponding electric field is $E_{\rm Coul} = e/r_0^2 \approx 1.9 \times 10^6$ CGSE (5.7×10^8 V/cm).

The potential barrier for electrons located near solid-vacuum interface equals the work function, which is lower than the above estimate. We describe the probability of electrons' penetration through the potential barrier, $\Delta_{\rm bar}$ (probability of ionization), by the tunnelling ionization of Keldysh (Keldysh, 1965):

$$w_{\rm tun}[s^{-1}] \approx \frac{2}{9\pi^2}\omega_{\rm bar} \times \left(\frac{E}{E_{\rm bar}}\right)^{5/2} \times \exp\left\{-\frac{\pi}{2}\frac{E_{\rm bar}}{E}\right\} \tag{5.2}$$

Here, $\omega_{\rm bar} = \frac{\Delta_{\rm bar}}{\hbar}$; $E_{\rm bar} = \Delta_{\rm bar}^{3/2} \times m_e^{1/2}/e\hbar$. The probability of the ionization above is averaged over the period much longer $2\pi/\omega$. Let us calculate with the help of (5.2) how many electrons might be ejected from the Al surface ($\Delta_{\rm bar}$ = 5 eV; $E_{\rm bar} = 1.354 \times 10^6$ CGSE) by the laser pulse $t_{\rm Ci}$ = 9.28 fs long with the average over the pulse intensity of 10^{14} W/cm^2 (field amplitude, 0.9×10^6 CGSE). With these parameters, one gets the ionization rate by the tunnelling, $w_{\rm tun} \approx 5.8 \times 10^{12}$ s^{-1}, suggesting that electrons from at least several atomic layers next to the surface are moved away leaving the positive ions being subject of the Coulomb explosion.

The non-relativistic equation governs the ions motion during the Coulomb explosion (Mourou et al., 2006):

$$\frac{d^2\xi}{dt^2} = \frac{\omega_{\rm pi}^2 \cdot r_0^3}{3\cdot(r_0 + \xi(t))^2} \tag{5.3}$$

Here, r_0 is the initial position of ion, and ξ is the displacement under the repelling force. The driving force above is the unscreened Coulomb force acting between ions with mass M, $F_{\rm Coul}^{\rm ii} = M\omega_{\rm pe}^2 \times r_0$ in plasma having the number density n at the average inter-atomic distance, $r_0 = \left(\frac{3}{4\pi n}\right)^{1/3}$. Equation (5.3) suggests that at the small displacements ($\xi << r_0$), the ion moves with the constant acceleration and in the opposite limit ($\xi >> r_0$) with the constant velocity. Thus, if the electron's energy exceeds the work function, $k_B T_e \geq w_{esc}$ to the end of the pulse of duration $t_p \leq \omega_{pi}^{-1}$, an electron escapes a metal leaving bare the parent ion. One can see that a 10 fs pulse with the intensity ~10^{14} W/cm^2 releases by the tunnelling the number of electrons, $n_e \approx w_{tun} \times n_a \times t_p = 3.48 \times 10^{21}$ cm^{-3}. The electron oscillating in such a field moves during a half-period time ($T/2$ = 1.33 fs for 800 nm light) from the parent ion on the distance of a few inter-atomic separations. Hence, the electrons from the couple of the mono-atomic layers next to the surface are removed from a metal.

The average electron energy to the end of the pulse at the threshold of ablation is expressed through the laser and optical parameters (absorption coefficient, A, and absorption depth, l_{abs}) of the laser-affected metal:

$$(k_B T_e)_{thr} = \frac{2A \times F_{thr}(t_p)}{C_e \times n_e \times l_{abs}} \tag{5.4}$$

Therefore, one can express the threshold fluence through the electron's temperature to the end of the pulse. However, the transient optical properties of the laser-excited metal at the conditions close to the ablation threshold are not measured experimentally. Therefore, the use of the estimate, $2A/l_{abs} \approx 4\pi/\lambda$, allows expressing the threshold fluence as follows:

$$F_{thr}(t_p) \approx \frac{C_e \times n_e \times \lambda_{abs}}{4\pi}(k_B T_e)_{thr} \tag{5.5}$$

The electron's energy at the threshold induced by the extremely short pulse (ESP) of duration $t_p \leq \omega_{pi}^{-1}$ should be $(k_B T_e)_{thr} \approx w_{esc}$ while for the longer pulses $t_p > \omega_{pi}^{-1}$, it is $(k_B T_e)_{thr} \approx w_{esc} + \varepsilon_b$. Thus, the threshold fluence for ablation by the ESP ($t_p \leq \omega_{pi}^{-1}$) is predicted to be $[w_{esc}/(\varepsilon_b + w_{esc})]$ times lower than by the conventional short pulses. For Al (ε_b = 3.065 eV; w_{esc} = 4.15 eV), this constitutes $[w_{esc}/(\varepsilon_b + w_{esc})]$ = 0.575. The above calculations implicitly assumed that ablation occurs in vacuum.

The threshold for the ablation of Al at 800 nm in vacuum by 15 fs pulse at 800 nm equals F_{thr} = 0.232 J/cm^2 (Genieys, et al., 2019). The threshold for the ablation of Al in vacuum by 12 ps pulse at 532 nm constitutes 0.32 J/cm^2 (Gamaly et al., 2005). The linear dependence of the ablation threshold on the laser wavelength is well established. Hence, at 800 nm, the threshold for Al is 0.48 J/cm^2 in agreement with the experiments. Therefore, for the ESP, the threshold is estimated as 0.575 × 0.48 = 0.276 J/cm^2 in a good agreement with the result of (Genieys et al., 2019) under assumption that 15 fs long pulse (at 800 nm) affects Al in the extremely short pulse mode.

The above formula (5.5) semi-quantitatively predicts the decrease of the threshold for the extremely short pulse (ESP). Another consequence of transition to ESP is the decrease of dependence on the laser wavelength due to the domination of the tunnelling mode for the electron removal, which is implicitly included in the estimation of the tunnelling ionization rate.

5.3.1.2 Comparison to the Coulomb explosions of clusters

It is instructive to compare the Coulomb explosion with the extremely short pulse described above to the Coulomb explosion of the clusters by the long pulses at the ultra-relativistic intensity (Mourou et al., 2006). At an intensity up to 10^{22} W/cm^2, electrons are stripped away from a cluster almost instantaneously by the ponderomotive potential (oscillation energy) not going through the secondary process of heating via collisions. In the case of the ESP, the oscillation energy is larger than the height of the barrier, $\varepsilon_{osc} > \Delta_{bar}$, making electrons tunnelling faster than ω^{-1} and not dragging the ions by the Coulomb force due to the brevity of the interaction, $t_p < \omega_{pi}^{-1}$. In a strong laser field (intensity > 10^{14} W/cm^2), the electrons acquire several electron-Volts of energy during the time less than half-period of the wave. Hence, the electrons move away to the distance equal to the of inter-atomic layer's thickness thus ensuring the electron/ion charge separation following by the Coulomb explosion. Note that there is a big difference between the Coulomb explosion of a small cluster and that from the surface of a massive target. Almost all electrons are irreversibly removed from a nanocluster forcing the whole cluster to explode. The electrons escaping from the surface of a massive target leave the positively charged ions in a single atomic layer (may be a few) at the threshold of ablation.

5.3.1.3 Ultra-short pulses $\omega_{\text{pi}}^{-1} < t_p < t_{\text{heat}}$

In order to consider the material removal by the laser action, three conservation laws—energy, momentum and mass—should be accounted for. In this section, we consider the pulses longer than the inverse ion plasma frequency ω_{pi}^{-1} but shorter than the energy transfer time to ions and the cooling time due to heat conduction. Hence, in the energy equation, the losses due to ions' heating and heat transfer by the electrons' diffusion can be neglected. The change of the energy during the pulse occurs due to laser absorption by electrons. The energy source term is expressed through the Poynting vector, boundary conditions, connecting the incident field to that inside a solid, and the transient optical properties of a material during the interaction (see Chapters 2 and 3). To the end of the sub-picosecond pulse, the whole absorbed energy density is confined entirely in the electrons sub-system. Integrating the energy equation over the pulse time, one expresses the electron's energy to the end of the pulse in the form of Eq. (5.4). We also use the estimate $2A/l_{\text{abs}} \approx 4\pi/\lambda$, reducing Eq. (5.4) to the following:

$$k_B T_e \approx \frac{4\pi \times F_{\text{th}}(t_p)}{C_e \times n_e \times \lambda} \tag{5.6}$$

Electrons are under the action of the laser field and the Coulomb forces between the charged particles during the pulse, $t_p > \omega_{\text{pi}}^{-1}$. The Coulomb collision rate in the conditions near the ablation threshold is around fs^{-1}. Therefore, part of the electrons oscillation energy in the high-frequency laser field is converted into the thermal energy via collisions and the condition, $T_e >> T_i$, holds. We are interested in the ions removal. This means that the characteristic time scale for the ion's removal is defined by the ion velocity, $v_i << v_e$, while the space scale should be larger than the inter-atomic distance. Hence, the electric potential changes slowly accompanied by the adiabatic adjustment of the electronic Boltzmann distribution function, $n_e = n_{0e} \times \exp\{e\varphi/k_B T_e\}$. Assuming that the quasi-neutrality holds, $n_e = n_i$, the force imposed by the hot electrons on cold ions through the electrostatic field reads (Lifshitz and Pitaevski, 1980) $eE_{\rm elst} = -k_B T_e \times \nabla \ln n_e$, (see Fig. 5.2). The momentum equation for the ion motion under the action of this field is the following:

$$M\frac{dv_i}{dt} = -k_B T_e \times \nabla \ln n_e \tag{5.7}$$

In order to move ion off a solid, the electron must have energy larger than the sum of the binding energy and the work function.

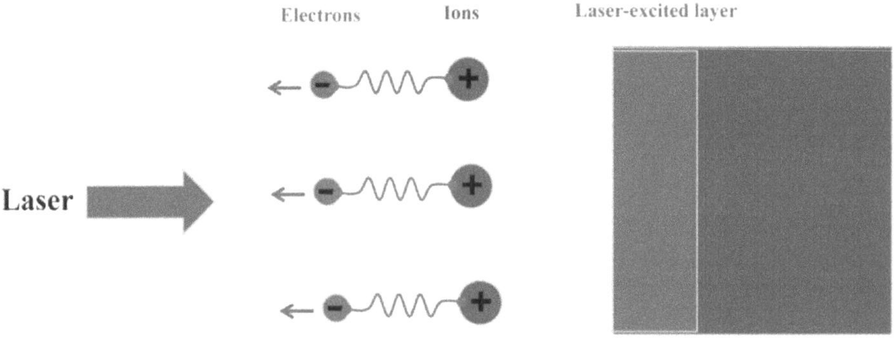

Figure 5.2 Electrostatic ("cold") ablation: energetic electrons drag the "*cold*" ions off the surface excited by the ultra-short pulse *before* the ions acquire the kinetic energy from electrons via the Coulomb collisions. Electrons are attached to ions by the Coulomb attraction (Gamaly et al., 2002).

The time to move an ion of mass M to the distance equal to the thickness of the atomic mono-layer from the surface, $r_{\rm mono}$ (ablate) is the following:

$$t_{\rm abl} \approx [2r_{\rm mono}/(dv_i/dt)]^{1/2} \tag{5.8}$$

Let us estimate the ion's acceleration from Eq. (5.7):

$$dv_i/dt \approx \{k_B T_e - (\varepsilon_b + w_{\rm esc})\}/M \times r_0 \tag{5.9}$$

At the electron energy in excess of a few electron-Volts over the sum of the binding energy and work function the Al ion moves during the hundred of fs (10^{-13} s) to the distance of a few inter-atomic layers from the surface indicating ablation.

5.3.1.4 Ablation threshold fluence for metals

The average electron's energy for ablation of metal should be larger than the sum of the work function plus the binding energy, $w_{esc} + \varepsilon_b$. The electron leaves the surface of the excited metal dragging behind it the much more massive ion. Taking absorption coefficient and absorption depth, electron number density and the heat capacity to the end of the pulse, we present the electron energy at the threshold of ablation from (5.4) in the following form:

$$k_B T_e = (w_{esc} + \varepsilon_b) = \frac{2A \times F_{th}(t_p)}{C_e \times n_e \times l_{abs}}$$

The threshold fluence for metals reads

$$F_{th}(t_p) = \frac{C_e \times n_e \times l_{abs} \times (w_{esc} + \varepsilon_b)}{2A}. \quad (5.10)$$

This formula has two distinctive features. First, it does not depend on the laser pulse duration because in the ultra-short interaction mode (pulse shorter than the energy transfer time to ions neither by collision nor by heat conduction), the electron's energy to the end of the pulse depends on the total energy density (fluence) delivered to a metal. Second, the ablation threshold by the ultra-short pulse (5.10) strongly depends on the laser wavelength (Gamaly et al., 2002). The electronic heat conduction dominates the interaction of the long (nanosecond) pulses that resulted in the strong dependence on the pulse duration while the dependence on the wavelength becomes weak.

The experimental results for the threshold fluence dependence on the pulse duration measured for Al ablated in air in the pulse duration range 15 fs ÷ 12 ps are presented in Fig. 5.3. In the figure, the thresholds vs. pulse time in a range of 0.1 ps ± 1 ns for gold ablation in vacuum are plotted (Stuart et al., 1996). In the pulse duration range 0.1–100 ps, the fluence is practically constant, 0.45 ± 0.1 J/cm^2. Change from the ultra-short regime to the long pulse regime occurs at around 100 ps and follows closely to the theoretical formula, $F_{th} = 0.049 \times (t_{ps})^{1/2}$ [J/cm²] (Gamaly et al., 2002). Experiments with other metals confirmed these two features of the ultra-short ablation regime. For example, the thresholds for ablation of Cu in vacuum by 532 nm laser beam were measured 0.375 J/cm² (9.6 ps) and 0.423 J/cm² (Nolte et al., 1997) and 0.41 ± 0.05 J/cm² (Gamaly et al., 2005).

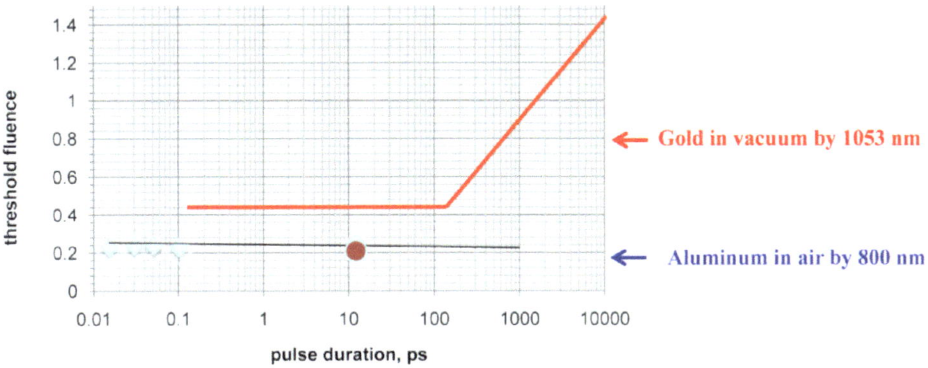

Figure 5.3 Ablation thresholds measured for gold (Stuart et al., 1996) and aluminium: blue squares (Genieys et al., 2019), red dot (Gamaly et al., 2005). Black line is a guide for eyes.

5.3.2 Ablation of Dielectrics

An atom at equilibrium can be removed from a solid (evaporated) if it gets energy in excess of the binding energy for breaking the inter-atomic bonds and getting enough the kinetic energy to leave a surface. In order to ablate the dielectrics, which are transparent at the low intensity, by the laser pulse, first, it is necessary to make them absorbing. Thus, the laser intensity should be high for the ionization, generating the necessary number of the electrons in the conduction band and converting the real part of the dielectric permittivity from positive to negative. At this interaction stage, a dielectric transforms into a metal-like state. The laser field exponentially decays inside the dielectric with the laser-modified optical properties at the length less than the laser wavelength (see Chapter 3). Hence, the first major difference of the laser ablation of dielectrics in comparison with that for metals is a need to deliver the energy per electron enough for moving an electron from the bound to the free state (in the order of the band gap, Δ_g). Besides, the free electron must have the energy in excess of the binding energy in order to drag an ion off the surface. Hence, from the energetic viewpoint, the ablation threshold for dielectric must be at least $(\Delta_g + \varepsilon_b)/(\varepsilon_b + w_{esc})$ times larger than for metal. The radical difference of the laser ablation of metals and dielectrics relates to the fact that at the threshold of metals ablation, the number of the conduction electrons is practically unchanged. In a dielectric excited close to the ablation threshold, the number of electrons in the conduction band changes by many of orders of magnitude approaching a metal-like state to the end of the pulse along with the drastic changes in the optical properties. In the laser-excited dielectric maintaining its solid state intact with the almost one conduction electron per atom to the end of the pulse, one can calculate the absorbed energy density in the surface layer in the same way as in the skin layer of metal:

$$E_{\text{abs}}(t_p) = \int_0^{t_p} \frac{2A(t) \times I(t)dt}{l_{\text{abs}}(t)}$$

The whole absorbed energy is in the thermal energy of electrons:

$$C_e \times n_e(t_p) \times k_B T_e = E_{\text{abs}}(t_p)$$

The threshold is defined by the condition that the electron's energy is equal to the sum of the band gap and the binding energy, $k_B T_e \approx \Delta g + \varepsilon_b$. Taking the optical properties, $2A/l_{\text{abs}}$, averaged by the pulse time (angular brackets in formula (5.11)) and introducing the fluence, $F(t_p) = \int_0^{t_p} I(t)dt$, one obtains the threshold fluence in the following form:

$$F_{\text{th}} \approx C_e \times n_e(t_p) \times (\Delta_g + \varepsilon_b)\left\langle \frac{l_{\text{abs}}}{2A} \right\rangle \tag{5.11}$$

In fact, this is a threshold for converting a solid into the solid density plasma with the inter-atomic bonds broken while the ions remaining at the initial temperature in the initial spatial positions.

One can define such a threshold expressed through the fundamental constants. Indeed, the electric field connecting electron to the ion core in the hydrogen atom at the first Bohr orbit, a_B, is $E_{\text{at}} = e/a_B^2$. The corresponding intensity of the laser field (plane polarized, averaged by the light periods) having such amplitude is $I_{\text{at}} = cE_{\text{at}}^2/8\pi$.

Introducing the atomic time scale $t_{\text{at}} = 2\hbar a_B/e^2$, one obtains the spatial energy density, $F_{\text{at}} = I_{\text{at}} \times t_{\text{at}} = c\hbar/2\pi a_B^3 = 2.0\ \text{J/cm}^2$.

The ablation threshold fluence has been measured and reported in numerous papers. Perry et al. (1999) measured the ablation threshold in fused silica for the pulse duration in a range from 100 fs up to 1 ns for the wavelength 1053, 825 nm and 526 nm. However, even the data related to the particular dielectric produced by the different groups is difficult to compare because the measurement methods and even the definition of the threshold are different. We present, for example, the set of results for the measurements of the ablation threshold for the fused silica (Table 5.2).

Table 5.2 Single-pulse ablation thresholds for fused silica versus pulse duration and laser wavelength

Pulse length	15 fs	25 fs	30 fs	100 fs	150 fs	300 fs	400 fs	10 ps	
SiO_2, 1053 nm, vac (Stuart et al., 1995)							2±0.3	~ 4	Long pulse: ≥ 10 ps
SiO_2, 1053 nm, vac (Stuart et al., 1996)					~1.8±0.3				Long pulse: ≥ 10 ps
SiO_2, 526 nm, vac (Stuart et al., 1995)							1 ± 0.15		transi-tion: ≥ 400 fs
SiO_2, 800 nm, vac (Tien et al., 1999)		3.0		~3.2			~3.2		transi-tion ≥ 400 fs
SiO_2, 800 nm, air (Chimier et al., 2011)	1.2		1.9	2.5		3.7			

It is seen from the table that even in the ultra-fast time domain, t < 400 fs, the data are scattered in a range of 1.2–3.7 J/cm^2. The data from Stuart et al. (1995) follows exactly the wavelength dependence, $F_{\text{th}} \propto \lambda$. However, the data from Tien et al. (1999) are in strong contradiction. Only in Tien et al. (1999), the threshold is practically constant from 25 fs up to 400 fs. In all measurements, the ultra-short regime lasts approximately up to 400 fs changing to the transitional regime, which holds up to 10 ps (Stuart et al., 1995; Tien et al., 1999). At the pulse durations, t_p > 10 ps, the long-pulse regime follows the familiar dependence, $F_{\text{th}} \propto t_p^{1/2}$.

Let us estimate the threshold for fused silica by formula (5.11) taking $\left\langle \frac{l_{\text{abs}}}{2A} \right\rangle \approx \lambda/4\pi$, $n_e \approx n_a = 6.6 \times 10^{22}$ cm^{-3}; $(J_1 + \varepsilon_b) = 11.79$ eV + 4 eV (averaged first ionization potential plus the binding energy), and $C_e = 3k_B/2$. One gets $F_{\text{th}}^{\text{Silica}} = 1.99 \times 10^4 \times (\lambda_{\text{cm}}) \left[\frac{\text{J}}{\text{cm}^2}\right]$. This estimate reproduces data of Stuart et al. (1995) for both wavelength, and underestimates data from Tien et al. (1999) and Chimier et al. (2011). No fitting coefficients were used in these estimates. We conclude that there is qualitative and semi-quantitative understanding of physics of the ablation mechanism of transparent dielectrics. Both estimates, one on the fundamental basis, and the other, which used the absorbed energy density in the surface layer of the laser-excited dielectric, give similar figures in qualitative agreement with the experiments.

The threshold measured for the shortest pulse of 15 fs (Chimier et al., 2011) is 1.58 times less than that for 30 fs pulse duration with no apparent physical reason. One may suggest that this measurement is an indication of transition to the extremely short pulse regime. Indeed, the inverse ion plasma frequency for Si and oxygen ions is in range of 15–13 fs. The average intensity per pulse is around 10^{14} W/cm^2 suggesting high ionization rate producing nearly one electron per ion. Oscillation energy at such intensity is around 6 eV per electron. The electron's collision frequency is near the natural maximum (5×10^{15} s^{-1}) suggesting that an electron can gain a few eV of energy during the pulse. Therefore, electrons can escape a solid surface moving onto the distance longer than the Coulomb attraction of the parent ions could return the electrons back after the end of the pulse. The charge balance is violated and the ions repulsion (the Coulomb explosion) blows the ions off the surface. Therefore, the laser must deliver to an electron the energy necessary for transition from the bound state to free state. The ablation threshold fluence for the extremily short pulse, $t_p \leq t_{C,I}$, should be less than that for the longer pulse (see Section 5.3.1.1). Similar estimates qualitatively agree with the observation (Chimier et al., 2011). The experiments in the referenced paper were performed in air. We show later that for such a short pulse the presence of air on the ablation threshold is insignificant.

5.4 Dependence of the Ablation Thresholds on the Laser Wavelength for Metals and Dielectrics

The absorbed energy density (and the threshold fluence) depends on the transient permittivity of the laser-excited solid through the following form:

$$\frac{l_{\text{abs}}}{2A} = \frac{\lambda}{8\pi} \frac{|1+\varepsilon^{1/2}|^2}{\varepsilon_{\text{im}}} = \frac{\lambda}{16\pi} \frac{[(n+1)^2 + k^2]}{n \times k} \tag{5.12}$$

Here, the wavelength dependence is expressed through the complex permittivity (or refractive index, $\varepsilon = \varepsilon_{\text{re}} + i\varepsilon_{im} = (n + ik)^2$. Note that in this derivation, the permittivity is taken in a most general form. The contribution of the valence electrons into the laser-excited dielectrics' permittivity should be accounted for. The threshold depends on the electrons' number density to the end of the pulse. For the laser-excited dielectrics, this term is also wavelength dependent.

$$F_{\text{th}} \approx C_e \times n_e(t_p) \times (J_1 + \varepsilon_b) \times \left\langle \frac{l_{\text{abs}}}{2A} \right\rangle$$

Metals. The measured wavelength dependence of the ablation thresholds for metals complies with the above linear formula.

Dielectrics. Comparison with experiments of Perry et al. (1999) and Stuart et al. (1996) for short pulse interaction with fused silica shows that in a wavelength range of 525–1053 nm, the thresholds follow the linear dependence.

5.5 Long Pulse Ablation Thresholds (Pulse Duration Longer than the Electron–Ion Energy Exchange Time and Heat Transfer Time)

In the ultra-short laser–matter interaction mode, the pulse duration is shorter of all relaxation times. The depth of the energy deposition zone near the surface in metals and in the ionized dielectrics is of the order, c/ω_{pe}, which is typically a few tens of a nanometer. In the interaction of the laser pulse of duration longer than the electron–ion energy transfer time that is normally of a few ps, the electronic heat conduction during the pulse increase the energy deposition region. Therefore, the length of the energy deposition region is the sum of the heat conduction length plus the absorption length, $l_{dep} = l_{th} + l_{abs}/2$. The heat conduction length is $l_{th} = (D \times t)^{1/2}$. The heat diffusion coefficient, D, in metals or in metal-like plasma is of the order of a few cm^2/s. Therefore, during a pulse a few ps long, the heat conduction depth is comparable to the absorption depth signifying the interaction mode transitional from the ultra-short interaction mode to the long pulse mode. This transitional mode is clearly seen in the experimentally measured dependence of the thresholds for metals (Stuart et al., 1995) and for dielectrics (Perry et al., 1999).

The energy conservation in one-dimensional case assumes that all deposited energy per unit area (absorbed fluence, $A \times F$) equals the thermal energy of a material at the depth, l_{dep}:

$$A \times F \approx C \times n_a \times k_B T \times l_{dep} \tag{5.13}$$

One can see that Eq. (5.13) in the limit $l_{dep} \approx l_{abs}/2$ and $T_i << T_e$ reduces to the ultra-short interaction threshold. The threshold for the long pulse ablation at $l_{dep} \approx l_{th}$, $T_i = T_e$, and $(k_B T \approx \varepsilon_b)$ reads

$$F_{\text{thr}}^{\text{long}} \approx C \times n_a \times \varepsilon_b (D \times t_p)^{1/2} / A. \tag{5.14}$$

This relation fits well with numerous experimental (Stuart et al., 1995, 1996; Perry et al., 1999) and theoretical studies with the nanosecond pulses (Afanasiev and Krokhin, 1971; Anisimov et al., 1971; Malvezzi et al., 1986; Anisimov and Luk'yanchuk (2002); Gamaly et al., 2002).

5.6 Ablation Thresholds in the Ambient Gas

Experiments revealed that the ablation thresholds for several metals (Cu, Al, Fe and Pb) in air are less than a half of those measured in vacuum. The single shots by the 12 ps (532 nm, 6.5 μJ) laser pulses the experimental conditions in air and in vacuum were identical: same focal spot, same energy per pulse. The thorough description of the experiments and their analysis are presented in Gamaly et al. (2005) and Gamaly (2011). The thresholds are presented in Table 5.3.

Table 5.3 Threshold fluence for ablation of metals by 12 ps (532 nm) pulses measured in air and in vacuum (Gamaly et al., 2005)

Metal, M (a.u.)	Al, 26.98	Cu, 63.54	Fe, 55.85	Pb, 207.19
F_{thr}, air (J/cm^2)	0.17±0.03	0.23±0.03	0.19±0.02	0.008±0.002
F_{thr}, vac. (J/cm^2)	0.32±0.04	0.41±0.05	0.36±0.04	0.08±0.02

The presence of air next to the laser-excited surface suffices the contribution of the equilibrium evaporation well after the end of the pulse. The succinct description of the effect is presented below.

The energy distribution in the outermost surface layer is responsible for the relative contribution of non-equilibrium ablation and thermal evaporation. Atoms from this layer are immediately leaving a solid when the energy per atom is in excess of the binding energy as described above.

In air adjacent to the laser-excited surface, however, collisions between the gas atoms and the surface result in establishing the high-energy tail in the energy distribution in the surface layer allowing thermal evaporation contribute even at the surface temperature below the binding energy.

After the laser pulse, the air next to the heated surface layer gains energy through collisions with the solid target. This results in the establishment of the Maxwell distribution in the air near the air–solid interface with the temperature in air equal to that in the laser-excited surface layer. Hence, the air plays the same role as the saturated vapour in the classical thermal evaporation. The presence of air introduces a new pathway allowing faster creation of the high-energy tail of the Maxwell distribution in the surface layer. Thus, two processes determine the ablation conditions at the solid–air interface: evolution of the Maxwell distribution at the surface due to air–solid collisions, and cooling of the surface layer by the heat conduction.

The presence of the air at $k_B \times T < \varepsilon_b$ reduces the time for the build-up of the high-energy tail in the surface layer allowing thermal evaporation to take place after the air–solid equilibrium has been established. The ablation rate then can be calculated using the thermodynamic phase equilibrium relations, which link the saturated vapour density (pressure) to the vapour temperature.

Let us first estimate the time for establishing the equilibrium distribution in air. The gas-kinetic mean free path in air in standard conditions equals $l_{g\text{-}k} = 6 \times 10^{-6}$ cm (Zel'dovich and Raizer, 2002), while the atomic velocity equals the sound velocity of $v_s = 3.3 \times 10^4$ cm/s. Therefore, the equilibrium distribution in air establishes well after the end of the pulse during $t_{\text{eq}} \approx l_{g\text{-}k}/\text{v}_\text{s} \approx 180$ ps. During this period, the laser-affected layer cools down due to the heat transfer to the cold region. Hence, the temperature in the distribution with the high-energy tail ($k_BT > \varepsilon_b$), responsible for the evaporation, drops from that to the end of the pulse, $T(t_p)$, to $T_{\text{eq}} = T(t_p) \times (t_p/t_{\text{eq}})^{1/2}$. The temperature at the end of the pulse is estimated using the fluence measured in the experiment and the absorbed energy density, $C \times n_a k_B T(t_p) = 2AF\,(t_p)/l_{\text{abs}}$.

Table 5.4 Maximum surface layer temperature at the ablation threshold fluence in vacuum and in air

	Al	Cu	Fe	Pb
$k_B T_m$ (eV), vacuum	2.5	2.9	2.39	1.15
$k_B T_m$ (eV), air	1.74	1.66	1.49	0.18
ε_b (eV)	3.065	3.125	3.965	1.795

Note: The binding energy is presented for comparison (Gamaly et al., 2005).

Now the contribution of the evaporation rate (atoms/cm^2) calculated as in the conventional thermodynamics integrating the atomic flow by time after establishing equilibrium:

$$\langle n \times v \times t \rangle_{\text{therm}} = \int_{t_{\text{eq}}}^{\infty} (n \times v)_{\text{therm}} dt \tag{5.15}$$

In accord with the suggestion that the air plays the same role as the saturated vapour in equilibrium, we take the air density as the saturated vapour density in the vapour's flow:

$$(n \times v)_{\text{therm}} \approx \frac{n_{\text{air}}}{(2\pi)^{1/2}} \left(\frac{k_B T_{\text{eq}}}{M} \right)^{1/2} \tag{5.16}$$

Inserting (5.16) to (5.15) and integrating, one can estimate the number of atoms, evaporated from the unit area:

$$\langle n \times v \times t \rangle_{\text{therm}} \approx \frac{n_{\text{air}}}{(2\pi)^{1/2}} \left(\frac{k_B T_{\text{eq}}}{M} \right)^{1/2} \times t_{\text{eq}} \left[\frac{\text{atoms}}{\text{cm}^2} \right] \tag{5.17}$$

The ablation threshold is defined as the absorbed energy density necessary to remove a single mono-atomic surface layer. Thus, the number of atoms evaporated above should be compared to the areal number density of atoms in a mono-atomic layer, $\langle n \times v \times t \rangle_{\text{therm}} \approx n_a \times d_a$.

Table 5.5 The predicted numbers of atoms thermally evaporated after the pulse once the Maxwell distribution has been established (it is compared with the number of atoms in a monolayer [Gamaly et al., 2005])

	Al	Cu	Fe	Pb
Thermal evaporation, × 10^{15} cm^{-2}	2.4	5.28	1.67	0.45
Areal density, × 10^{15} cm^{-2}	1.72	2.16	2.0	1.15

Table 5.5 suggests that thermal evaporation well after the end of the laser pulse in air can, indeed, be responsible for the removal of a mono-atomic layer for Al, Cu and Fe in a semi-quantitative agreement with the experiments (Gamaly et al., 2005). Therefore, the presence of air *decreases* the single-pulse ablation threshold by a factor of 2 relative to the vacuum case due to the contribution of thermal ablation assisted by the presence of the air well after the end of the pulse.

Recently, the ablation thresholds of aluminium in air were measured versus pulse duration from 15 to 300 fs by 800 nm pulses (Genieys et al., 2019) being almost pulse time independent in a range, 0.229–0.240 J/cm^2. One can compare the result of Gamaly et al. (2005) at 532 nm with that from Genieys et al. (2019) at 800 nm for Al taking the threshold fluence for metals proportional to the laser wavelength, $F_{thr} \sim \lambda$. Then from Table 5.3 one obtains F_{th} (800nm) = 0.255 J/cm^2 close to results (Genieys et al., 2019).

5.7 Ablation Rate, Depth and Mass Per Single Pulse

For the interpretation and prediction of the ablation experiments, one needs simple formulae resulting from the basic conservation laws. Let us consider a single pulse delivering energy per unit surface (fluence, J/cm^2) in excess of the ablation threshold. The diameter of the focal spot is much larger than the ablation depth. Thus, the one-dimensional approximation is legitimate. An upper limit for the ablation depth then follows from the energy conservation law. The binding energy per atom, ε_b, should be spent to free an atom (losses neglected). Suggesting that all absorbed energy is spent on the material removal, one obtains the maximum energetically possible ablation length:

$$l_{\mathrm{abl,max}} \approx d_{\mathrm{mono}} + \frac{A(F - F_{\mathrm{thr}})}{n_a \times \varepsilon_b} \tag{5.18}$$

Here, $d_{\mathrm{mono}} \approx n_a^{-1/3}$ is the thickness of the atomic monolayer, A is absorption coefficient and F_{thr} is the threshold fluence. The losses for the kinetic energy of ablating atoms and for the heat losses should be added the F_{th} into parentheses in the denominator for a more realistic estimate.

The ablation depth of metals, Al, Cu, Fe and Pb, measured as a function of the increasing fluence grows faster than the linear dependence of (5.18) (Gamaly et al., 2005). The experimentally measured functions, l_{abl} (F), approach the idealistic formula (5.18) at the elevated fluence. The experiments with ablation of fused silica in a range of 5 fs to 5 ps by Lenzner et al. (1999) demonstrated that the ablated volume ($V_{\mathrm{abl}} \approx S_{\mathrm{foc}} \times l_{\mathrm{abl}}$) grows up in the direct proportion to fluence in agreement with (5.18).

The ablated mass is connected to the ablation length, laser focal spot area, S_{foc}, atomic number density and atomic mass, M, by the obvious relation:

$$m_{\mathrm{abl}} = S_{\mathrm{foc}} \cdot n_a \cdot M \cdot l_{\mathrm{abl}} \tag{5.19}$$

The ablation mass per single pulse can be measured when the high-repetition-rate laser pulse scanned over the sample surface with a speed separating single-pulse action in space and time. Then ablated mass per single pulse is measured averaged by thousands of the identical events (Gamaly et al., 2005). The ablation depth per single pulse is then recovered with the help of Eq. (5.19).

Thermal ablation. Ablation, produced by the ultra-short pulses, proceeds in two stages at the fluence in excess of the threshold: In the first stage during the pulse, when the ions are still cold, non-equilibrium (electrostatic ablation) takes place. The electronic heat transport begins during the pulse becoming dominant after the end of the pulse. After the end of the pulse, the electron temperature decreases due to the energy transport by the electronic heat conduction. The electronic heat wave starts to propagate inside a cold material with the temperature-dependent velocity decreasing during the wave propagation. When the thermal wave delivers to ions the energy in excess of the binding energy, the ions are detached and move away (ablated) with the sound velocity corresponding to the local temperature behind the heat front. Thus, the thermal ablation process induced by the heat wave can be presented as the non-linear heat wave front propagated inside a material followed by the ablation front moving with the local speed of sound that is less than the heat wave speed at the beginning of the process. The ablation terminates when the ablation front catches up with the heat wave front (Afanasiev et al., 1999). At this moment, electrons and ions temperatures are equilibrated and the shock wave is generated. The scheme of the thermal ablation process is presented in Fig. 5.4.

It is possible to derive the scaling relations expressing the heat conduction (ablation) depth and ablation time through the laser and material parameters. The energy conservation and electronic heat transfer control the process. The ablation process can be considered as one-dimensional because the energy of the absorbed fluence converted to the energy in the absorption layer, the thickness of which is much less than the focal spot diameter. During the heat wave propagation, the energy is confined between the surface of the sample and the heat wave front with the coordinate, x_f:

$$F = C_e n_e k_B \int_0^x T dx \approx C_e n_e k_B T \cdot x_f \tag{5.20}$$

Subscript "*f*" marks values at the heat front while subscript "0" relates to the initial values. The second equation for the scaling is the non-linear heat conduction equation:

$$\frac{\partial T_e}{\partial t} = D \frac{\partial^2 T}{\partial x^2}$$

The diffusion coefficient for the ideal plasma has the form $D = D_0(T_f/T_0)^{5/2}$. We present this equation in a scaling form as follows:

$$D_0\left(T_f/T_0\right)^{5/2} \approx x_f^2/t \tag{5.21}$$

It is convenient to present both equations in the dimensionless form introducing dimensionless variables:

$$\xi = \frac{x_f}{x_0};\ \theta = \frac{T_f}{T_0};\ \tau = \frac{t}{t_0}$$

Then Eqs. (5.20–5.21) take the following form:

$$\xi^2 = \delta_0\theta^{5/2} \times \tau;\ \delta_0 = \frac{D_0 t_0}{x_0^2};\ \theta \times \xi = \varphi_0;\ \varphi_0 = F/C_e k_B T_0 n_e x_0 \tag{5.22}$$

The heat front temperature is expressed as the function of time and the heat front position. The dimensionless heat front velocity follows from (5.22):

$$v_{\text{front}}\left(t_0/x_0\right) = \frac{d\xi}{d\tau} = \frac{2}{9}\delta_0^{2/9}\varphi_0^{5/9}\tau^{-7/9} = \frac{2\delta_0\theta^{7/2}}{9\varphi_0} \tag{5.23}$$

The ablation velocity equals the local sound velocity behind the heat front, $v_{\text{abl}} = (3k_BT/2M)^{1/2}$. We suggest that the ablation terminates at the moment, when the sound velocity reaches the heat front velocity, $v_{\text{abl}} = v_{\text{front}}$. Note that this condition also signifies the electron–ion temperature equilibration and transformation of the heat wave into the shock wave at the sufficiently high fluence. This equality allows the determination of the temperature and ablation depth as functions of the laser and plasma parameters at the moment when the ablation terminates:

$$T_{\text{abl}} = T_0\left(\frac{9v_0t_0}{2x_0}\right)^{1/3}\left(\frac{\varphi_0}{\delta_0}\right)^{1/3} \propto \left(\frac{F}{D_0}\right)^{1/3} \tag{5.24}$$

Using (5.22), one obtains the ablation depth, $l_{\rm abl} \equiv x_f(t_{\rm abl})$, and the ablation time, $t_{\rm abl} = t_0\varphi_0^2/\delta_0\theta_{\rm abl}^{9/2}$:

$$l_{\rm abl} = x_0\left(\frac{2x_0}{9v_0t_0}\right)^{1/3}\varphi_0^{2/3}\delta_0^{1/3} \propto F^{2/3}D_0^{1/3}$$

$$t_{\rm abl} = \frac{t_0\varphi_0^2}{\delta_0\theta_{\rm abl}^{9/2}} = t_0\left(\frac{2x_0}{9v_0t_0}\right)^{3/2}\varphi_0^{1/2}\delta_0^{1/2} \propto F^{1/2}D_0^{1/2} \tag{5.25}$$

The scaling for the non-equilibrium and thermal ablation stages gives semi-quantitative agreement with the experimental data and reveals the physics of ablation. The non-thermal ablation is more effective than the equilibrium ablation: Ablation depth is directly proportional to the laser fluence, $l_{a,n\text{-th}} \sim F$, in agreement with the experiments for ablation of metals (Gamaly et al., 2005). The thermal ablation occurs after the end of the pulse, while the losses due to cooling by heat transfer are essential. Therefore, the thermal ablation is less efficient, $l_{a,\rm th} \sim F^{2/3}$. The scheme of thermal ablation is presented in Fig. 5.4.

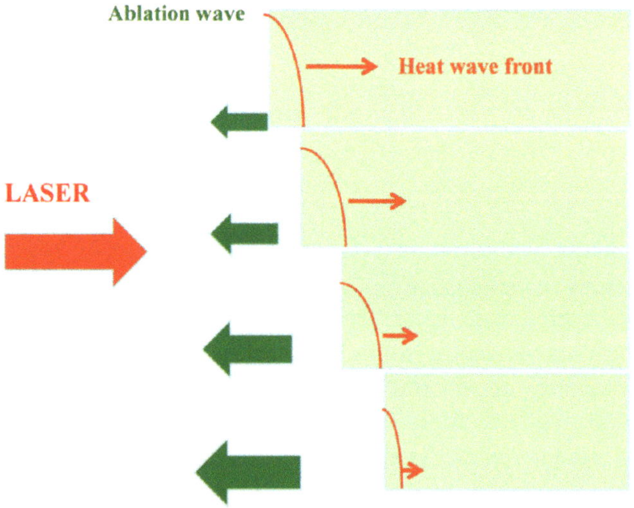

Figure 5.4 Thermal ablation: Electronic heat wave propagates inside a solid until the sound velocity behind the front equals the heat front velocity.

It is instructive to notice the typical ablation parameters per single 100 fs pulse at the fluences a few times over the ablation threshold: ablation depth ~100 nm; ablation time ~ a few picoseconds. Ablation terminates well after the end of the ultra-short pulse.

5.8 Control over the Phase State of the Ablated Plume

The laser-ablated plasma is used for the deposition of thin films of different materials. In order to produce a film with the atomic surface finish and homogeneity, the plasma plume arriving on the substrate should be homogeneous and atomized. There are two major reasons for the plume to have a complex composition. The first relates to the spatial intensity variations across the laser focal spot. The onset of the processes transforming the state of a target under the action of the intense laser pulse depends on the local laser intensity/fluence. These processes appearing progressively as the laser intensity increases are the structural phase transitions, destruction of the target integrity: appearance of cracks, flaking of the surface, melting, ablation, and ionization. The spatial intensity distribution across the focal spot usually has a Gauss-like form. Therefore, the whole variety of the above-mentioned processes takes place across the laser-affected area even if the average spatial intensity is higher than the ablation threshold. A plume produced by a laser with such intensity variations can contain material in different phase states: gas, melt, etc.

Another source of the inhomogeneity, in particular the formation of liquid droplets in the plume, is condensation of the vapour during the expansion. As a result, an expanding plume consists of a mixture of gas and liquid phases. Therefore, in order to ensure the homogeneity of a plume, several conditions should be fulfilled: (1) The local fluence at every point across the focal area should exceed the ablation threshold. (2) The gas density and the temperature of the expanding plume should satisfy the thermodynamic relations corresponding to the gas phase being above the condensation point. (3) The spatial intensity distribution across the focal spot should be all the same.

The threshold energy fluence, $F_{thr} \propto H_{transf}$, scales with the characteristic energy required for the particular phase transition in equilibrium. For example, the ratio of the energy for vaporization, H_{vap}, to the enthalpy of fusion (melting), H_{melt}, is proportional to the ratio of the ablation to the melting threshold. This ratio varies for most materials within a range ~ 5–30. For example, for iron, H_{vap}/H_{melt} = (340 kJ/mole/13.89 kJ/mole) = 24.6. At the periphery of the focal spot having the Gauss distribution for the fluence across the spot, the phase state of the laser-affected material is completely different for that in the centre.

Therefore, the energy in excess of the phase transition threshold should be deposited evenly over the focal spot area in order to make homogeneous the phase-state of the plume. This condition emphasizes that the threshold energy should be significantly higher than that required to just break the inter-atomic bonds. Sufficient kinetic energy should be additionally delivered to an unbound atom in order to remove it from the solid and maintain it in the desired vapour state. The production of a droplet-free laser plume imposes the additional condition that the gas state of the vapour should be effectively collisionless.

5.8.1 Criterion for Complete Atomization of Ablated Plume

The energy threshold for the total atomization of the ablated plume can be estimated on the basis of the equilibrium thermodynamics similar to the criterion for complete vaporization of material under unloading after the arriving the strong shock wave from the bulk at the solid/vacuum interface (Zel'dovich and Raizer, 2002). The deposited laser energy should be sufficient for the breaking inter-atomic bonds and for the supply of additional kinetic energy necessary for keeping the expanding plume in the atomized state. The absorbed energy density expressed through the laser parameters then determines the threshold for target atomization.

The pressure–volume (P, $V = n^{-1}$) or temperature–volume (T,V) charts define at the phase plane the sets of the parameters where the phase states for a particular material can exist.

The phase equilibrium curve at the phase plane separates the liquid, solid and gas states from the states with a mixture of phases (see Fig. 5.5). The maximum of the phase equilibrium curve in the temperature/density plane is the critical point. The phase states of an expanding gas, continuing to be in atomized state, characterized by T,n points on the isentropic curve along which the entropy of gas is constant. An expansion obeying the condition on the isentropic curve coined by the adiabatic expansion. The atoms in an adiabatic expansion do not interact with each other (cannot form clusters or droplets). It is obvious that all adiabatic curves should lie above the critical point.

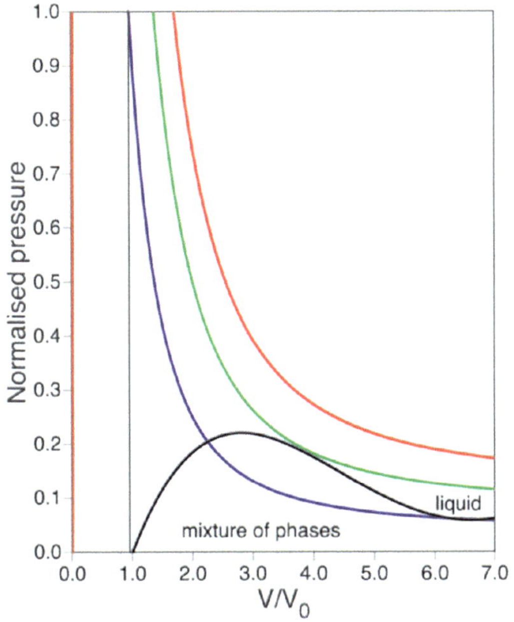

Figure 5.5 *P–V* diagram of the vapour states in the plume at various absorbed laser energy (different initial pressure). The total atomization is achieved at the conditions when the *P–V* curve touches the critical point (green line). The black curve of the phase equilibrium following from the Clapeyron–Clausius equation separates the states with the mixture of phases and the gas phase.

The equation of state (relation between pressure, density and temperature) of the ablated material in conditions close to that for the solid–vapour phase transition can be presented as a sum of the elastic (cold) pressure (related exclusively to the inter-atomic interaction) and the thermal pressure, $P = P_{\text{th}} + P_e = P_{\text{th}} - |P_e|$. The material in the laser-excited layer is swiftly elevated to the high temperature under the action of a short powerful laser pulse while the initial density remains unchanged. Hence, the initial state of expansion is characterized by the temperature after the electron and ion equilibration, T_{eq}, and the initial density of expanding solid, n_0. A simplified approach allowing catch up the main idea is considering the ablated plume as the ideal gas during the whole expansion. The isentrope (function signifying the entropy conservation) for the ideal gas has a form, $k_B T/n^{2/3}$ = const. The expansion of a plume remaining in the atomized state should be during expansion in states located above the critical point. The limit case of the isentrope touching the critical point obeys the following condition:

$$k_B T_{\text{eq}} \Big/ n_0^{2/3} = k_B T_{\text{cr}} \Big/ n_{\text{cr}}^{2/3} \qquad (5.26)$$

The states above the first adiabatic curve touching the critical point represent the atomic state of a homogeneous phase at ($n < n_{\text{cr}}$). The adiabatic curve, which separates the single-phase area from the mixture of phases, and the phase equilibrium curve have only one common point, e.g., the critical point. Therefore, if it would be possible to express the critical values, T_{cr}, n_{cr}, through the basic characteristics of a solid, then the temperature at the beginning of expansion (end of the pulse) would be also expressed through the basics. The critical values are poorly known for many materials.

The critical parameters, T_{cr}, n_{cr}, at least in principle, can be related to the binding energy and the initial density (Zel'dovich and Raizer, 2002; Landau and Lifshitz, 1980) as follows: $T_{\text{cr}} \approx B_1 \times \varepsilon_b$; $n_0/n_{\text{cr}} \approx B_2$. The numerical coefficients can be found using the available experimental and computational data.

For example, using the experimental data for the critical pressure for aluminium and interpolation for the "cold" pressure (More et al., 1988), one gets $n_0/n_{cr} \approx 26$, and the critical pressure $P_{cr} \approx 5.55 \times 10^{-3}\ n_0 \times \varepsilon_b = 1.84$ kbar. On this basis, the adiabatic expansion curve for aluminium touching the critical point must start at the end of the laser pulse after electron to ion temperature equilibration at $k_B T_{eq} \approx 4.9\varepsilon_b$. From this calculation, the practical conclusion follows: In order to obtain the atomized plume of ablated aluminium, the laser fluence should be a few times larger than the threshold fluence, $F > 5F_{thr}$. The binding energy for the majority of materials is in the range 3–5 eV per atom. Therefore, the above criterion for the plume atomization is applicable for different materials.

Semi-quantitative, however physically well substantiated, conclusion is the following. For obtaining the fully atomized plume a solid should be overheated to the level when the ablated gas expands adiabatically at the temperature well over the critical point. The intensity (fluence) spatial distribution across the focal spot should be homogeneous ("flat top hat" distribution).

5.8.2 Experimental Verification of the Plume's Atomization

Two processes affect the ablated plume atomization. First, the fluence at a particular laser-affected spatial spot should be a few times higher than the ablation threshold. However, in the conventional Gauss spatial distribution of the intensity across the focal spot, the low-intensity wings inevitably present at the periphery of the spot produce the droplets of melted material. Therefore, two conditions should be imposed on the laser intensity spatial distribution across the focal spot: The intensity should be spatially homogeneous, and fluence at any space point must be a few times higher than the ablation threshold.

A relatively simple, however indirect, method employed for checking the plume's atomization is the following. The expanded laser-ablated plume condenses on a substrate placed close to the focal spot forming thin film of ablated material. The quality of deposited film, which can be thoroughly examined under microscope (optical, electronic, AFM), is indirect evidence of the plume's composition. The quality of the silicon films deposited by the plume generated by the laser pulses with different spatial distribution of intensity over the laser focal spot and with the different fluence was examined in Gamaly et al. (2004). The fluence for the total atomization of the silicon plume is 4 times the ablation threshold.

Three series of the deposition experiments were performed using a 150 fs laser. (a) In the first series, the target was located at the focal spot (Gauss distribution, maximum fluence higher than the atomization fluence). (b) In the second series, the target was located at the image plane of the 2.2 mm diameter aperture. (c) In the third series, the target locates at the same position but with the aperture removed. Introduction of the aperture practically removes the low-intensity wings in the Gauss distribution of the intensity over the focal spot making it close to the "flat-top-hat" one. The examination of the films deposited with the truncated wings and with the laser fluence 4 times over the ablation threshold revealed the best quality: The density of droplets was reduced below 10 per cm^2. This experiment demonstrated that elimination of the low-intensity wings in the spatial distribution at the high average fluence resulted in the formation of the atomized plume. Following the adiabatic expansion of the ablated plume and its deposition on the substrate led to the formation of the good quality of the deposited film surface.

5.9 Multiple Pulse Action: Accumulation Effects

It was demonstrated that single short intense pulse evaporates 10^{11}–10^{12} atoms per pulse (Rode et al., 1999; Perry et al., 1999). However, the high average ablation rate per second is necessary for the industrial applications of laser ablation. To compensate for the low ablated mass per pulse, trains of millions pulses per second are then used to achieve a high ablation rate per second.

The interaction of the pulse trains from MHz repetition rate lasers with matter appears to be significantly different from the single-pulse interaction due to the coupling of the effects of the successive pulses. The high repetition rate is accompanied by the high scanning speed of the laser beam across the interaction surface allowing varying regimes from the single-pulse action on the focal spot to the regime when up to thousands pulses arrive at the same spot. The energy accumulation due to the coupling between the low-energy consecutive pulses gradually increases the energy density over the ablation threshold.

The mechanisms for the coupling of successive pulses in interactions with metals and transparent dielectrics with low heat conduction (such as silica and glasses) are different. The cumulative ablation of metals occurs due to the gradual plume density and temperature increase in the vicinity of the target surface. This leads to a change in the laser–matter interaction mode from the absorption at the solid surface to absorption on the plasma density gradient of the expanding plume.

In the interaction with transparent dielectrics, the time between successive pulses is too short for the completion of the cooling of an irradiated spot. Thus, the energy accumulation within the target is the major effect in the cumulative ablation of poorly absorbing dielectrics. As a result, high-repetition-rate pulse trains allow the precise control over the temperature at the sample surface via the "dwell" time of many pulses at the same spot, allowing ablation to occur under conditions when the single-pulse fluence is insufficient to ablate the material.

The cumulative effect in the high-repetition-rate-laser interaction with matter has been observed during the ablation of carbon by a laser with 76 MHz repetition rate (Rode et al., 1999). The effects of high-repetition-rate ablation, which include the smoothing of the spatial intensity distribution and cumulative heating, have both been observed in ablation and deposition of chalcogenide glasses (Rode et al., 2002; Rode et al., 2001). Cumulative heating in the bulk of transparent glass was also reported in Schaffer et al. (2003).

In this section, we will consider briefly experiments on ablation of aluminium, copper, steel, lead and silicon targets with a 4.1 MHz pulse-rate 10 ps mode-locked laser in air and in vacuum at conditions above and below the single-pulse threshold. The analysis of these experiments allows the introduction of the specific parameters of the multiple pulse action and reveals its physical meaning and role in controlling the interaction process.

5.9.1 Dwell Time: Control over the Number of Pulses Hitting the Spot

Two new parameters, scanning speed, v_{scan}, and repetition rate, R_{rep}, are needed for the control over the interaction process changing the number of pulses hitting the focal spot. It is convenient to introduce a time period for how long the laser beam "dwells" over a focal spot of a diameter, d_f, for a given scanning speed, v_{scan}, $t_{dw} = d_f/v_{scan}$. Then the number of pulses hitting the same spot at the repetition rate, R_{rep}, is as follows:

$$N_{foc} = \frac{R_{rep} \times d_f}{v_{scan}} \tag{5.27}$$

In experiments where the laser beam was scanned with the oscillating mirrors at 76 MHz repetition rate laser (R_{rep} = 76 MHz; d_f = 25 μm; $(v_{scan})_{av} \approx$ 50 mm × 100 Hz = 5 m/s), the number of pulses per spot varied in a range from 38 to 7 × 10^4 (Rode et al., 1999). Adjusting repetition rate and scanning speed, one can choose either the single-pulse interaction mode or the multiple pulse modes where different steering levers changing the interaction are available.

5.9.2 Multiple-Pulse Action on Dielectrics: Energy Accumulation

Let us consider the multiple pulses hitting the surface of the low absorbing poor-heat-conducting dielectrics such as silica and glasses. The effect of the focal spot cooling during the time gap between the consecutive pulses is important for the energy accumulation. The first pulse heats the spot to the temperature T_1. The heat conduction can be treated as one-dimensional process because the heat conduction length, l_{th}, and absorption length, l_{abs}, are both much less than the focal spot diameter, $l_{th} << l_{abs} << d_{foc}$. The characteristic cooling time, t_{th}, of the absorption layer with thickness l_{abs} comprises $t_{th} = l_{abs}^2/D$. D is the thermal diffusivity in cm^2/s. The time gap between the successive pulses is inversely proportional to the repetition rate, $\Delta t_{gap} = R_{rep}^{-1}$. Therefore, we suggest that the temperature drops after the end of the first pulse in accordance with the one-dimensional heat conduction as follows:

$$T \approx T_1 \left(\frac{t_{\mathrm{th}}}{t_{\mathrm{th}} + R_{\mathrm{rep}}^{-1}} \right)^{1/2} \equiv \alpha T_1 \tag{5.28}$$

The condition $R_{\mathrm{rep}}^{-1} << t_{\mathrm{th}}$ holds even at 100 MHz repetition rate. Therefore, $\alpha \le 1$, and the temperature drop after a single-pulse action simplifies to the following form:

$$\alpha \approx \left(1 - \frac{R_{\mathrm{rep}}^{-1}}{2t_{\mathrm{th}}} \right) \tag{5.29}$$

The temperature accumulation after the N-th pulse hitting the same spot then reads

$$T_N = T_1(1 + \alpha + \alpha^3 + \cdots + \alpha^N) = T_1 \frac{1-\alpha^N}{1-\alpha}. \tag{5.30}$$

The effect of the temperature accumulation is very strong in chalcogenide glass where the both heat conduction and absorption are very low: $D = 1.15 \times 10^{-3}$ cm^2/s; $l_{\mathrm{abs}} = 2 \times 10^{-4}$ cm; $t_{\mathrm{th}} = 3.48 \times 10^{-5}$ s. In these conditions, the temperature in the absorption layer grows up as $T_N \approx NT_1$ until the thermal ablation threshold can be reached (Rode et al., 2002; Rode et al., 2001).

5.9.3 Achieving Thermal Ablation Threshold with the Low-Energy Multiple Pulses

Strong accumulation effect was observed in the experiments on the ablation of chalcogenide glass by the laser with 76 MHz repetition rate (Luther-Davies et al., 2005). The accumulation effect is strong due to big number of pulses hitting the focal spot. The temperature accumulated after the action of N pulses is $k_B T_N = 293K + N \cdot k_B \cdot \Delta T_1$ (ΔT_1 is the temperature increase after the single-pulse action). The number of pulses hitting the spot necessary for achieving the thermal ablation threshold can be determined using the formula for the thermal ablation threshold defined above:

$$\left(\frac{2k_B T_N}{M}\right)^{1/2} \cdot \exp\left\{-\frac{\varepsilon_b}{k_B T_N}\right\} \cdot R_{\text{rep}}^{-1} \approx d_a \tag{5.31}$$

The solution to this equation in accord with the experimental conditions suggests that the ablation threshold was achieved during the time gap between 91th and 92th pulses. Important conclusion from these experiments and analysis is that the additional steering parameters, repetition rate and scanning speed, allow achieving the efficient ablation when single-pulse energy is insufficient for ablation (Luther-Davies et al., 2005).

5.9.4 Smoothing of the Intensity Distribution across the Focal Spot

There is an interesting effect in the multiple pulses action connected with the interplay between the high repetition pulse rate and the beam scanning speed. At the high repetition rate the distance separated two consecutive laser beams scanned along the focal spot might be much less than the focal spot diameter, $\Delta x = v_{\text{scan}} \times R^{-1} << d_f$. Single laser pulse focused at a target surface produces a Gaussian intensity distribution across the focal spot. However, $N = d_f/\Delta x >> 1$, pulses moving across the focal spot produce almost homogeneous intensity distribution across the spot. If the maximum intensity of a single pulse only slightly exceeds the ablation threshold, then the plume produced by such a pulse contains a mixture of phases: liquid droplets, flakes, etc. The spatial intensity distribution must be modified in order to control the phase state of the ablated vapours as demonstrated earlier.

The use of a high repetition rate is, however, a natural way to produce a plume of homogeneous phase. The spatial intensity distributions are overlapping at the focal spot during the dwell time making the intensity distribution over the focus and therefore the evaporation conditions almost homogeneous. All the imperfections of a single beam are smoothed with the successive pulses heating the same surface spot several times. As a result, it may be unnecessary to control the intensity distribution as required for single-pulse irradiation.

5.9.5 Change of the Interaction Mode: Density Build-Up near the Target Surface

The presence of air next to the solid surface helps faster establishment of the Maxwell distribution and increases the ablation rate due to contribution of the thermal evaporation. Multiple pulses hitting the target surface get a similar effect due to the accumulation of a dense vapour in front of the solid target surface. The increase in the vapour density is equivalent to the increase in the saturated vapour density, and therefore it results in an increase in the thermal ablation rate.

Let us estimate the conditions for the density build-up making this interaction mode efficient. First, the Maxwell distribution in the vapour next to solid should be established during the time shorter than the time gap between the pulses, $t_{eq} = (n_{vap}\sigma v_{th})^{-1} < R_{rep}^{-1}$. Hence, the vapour density should comply with the inequality $n_{vap} > R_{rep}/\sigma v_{th}$. The experiments and analysis from Luther-Davies et al. (2005) suggest that at the repetition rate 4.1 MHz ($\sigma \sim 10^{-15}$ cm^2; and $v_{th} \sim 10^5$ cm/s), the vapour density should increase more than 200 times (from $n = 1.8 \times 10^{14}$ cm^{-3} at $P = 3 \times 10^{-3}$ Torr to $n > 4 \times 10^{16}$ cm^{-3}).

Let us consider the conditions for such density build-up. The plume expands adiabatically because the entropy and the energy are both conserved after the end of the pulse. The density and the temperature of a plume during the isentropic expansion go to zero at the finite distance from the initial position, while the velocity is at maximum (Zel'dovich and Raizer, 2002). Therefore, the density next to ablation surface has a steep gradient. The size of the expanding plume grows linearly with time (and with the number of consecutive pulses hitting the same spot), $x_N \leq N \cdot v_{th}/R_{rep}$.

Analysis of silicon ablation by 4.1 MHz laser along with the similar analysis of the chalcogenide glass ablation by 76 MHz laser demonstrates that around hundreds of pulses per spot is necessary to build up the dense plume next to the target surface dense enough to "switch on" thermal evaporation in the manner invoked in the presence of air.

The interaction regime begins changing to interaction with the plasma density gradient. The absorption occurs at the critical density significantly increasing the temperature in the absorption zone and near the ablation surface. Efficient ablation continues during the subsequent several hundred pulses after the laser–plasma interaction regime has been established. The scan rate in these experiments allowed several thousand pulses to hit the same spot (Luther-Davies et al., 2005).

5.10 Summary

Evaporation of a solid by any means occurs if the atoms get the energy in excess of the binding energy. Ablation by the lasers with the pulse duration shorter of major relaxation times occurs in non-equilibrium conditions when the electrons get the entire absorbed energy at the end of the pulse while ions remain in the initial cold state during the pulse. There are specific pathways for the coupling of the energetic electrons to cold ions leading to the ions removal from laser-excited solid depending on the laser pulse duration:

- Under the intense pulse shorter than the inverse ion's plasma frequency, $t_p < \omega_{\mathrm{pi}}^{-1}$, energetic electrons escape a solid surface faster than the Coulomb force may affect ions. The Coulomb repulsion of the bare positive charges removes the ions from the surface (ablation by the Coulomb explosion)
- During the longer pulses, $\omega_{\mathrm{pi}}^{-1} < t_p < t_{ei}, t_{\mathrm{th}}$ but shorter of the time for the electron–ion energy transfer by collisions and heat conduction the energetic electrons getting enough energy to escape a solid and drag off the cold parent ion attached to electron by the electrostatic attraction ("cold" ablation). Ablation threshold for such pulses is independent on the pulse duration.
- For pulses longer than the electron-to-ion energy transfer time and heat conduction time, the ionic energy distribution

is nearly Maxwell and ablation proceeds in the thermal equilibrium mode. The ablation threshold for the long pulses depends on the pulse duration, $F_{\text{thr}} \propto t_p^{1/2}$.

- Using the high-repetition-rate laser beam scanning along the surface of the sample, one can change the interaction mode exploring the effects of energy accumulation significantly increasing the efficiency of ablation (grams/sec).

Chapter 6

Extreme Energy Density Confined inside a Transparent Crystal—Novel Path for New Materials Creation: Solid-Plasma-Solid Transformations

6.1 Introduction

There has been a century-long quest to create high-pressure/-temperature conditions in laboratory, study the behaviour of materials under extreme conditions, and produce new materials with unusual properties. The invention and improvements of the diamond anvil cell during the past century resulted in obtaining the static pressure approaching to TPa with simultaneous heating by laser up to 3,000 K (McMillan, 2002, 2005). Recently the conditions at the Earth's core of 364 GPa and 5,500°C were recreated in the lab (Tateno et al., 2010) The diamond anvil method allows the study of the process of the phase transformation and preserves the novel phase ready for the further studies. However, the upper limit for the high pressure is set by the strength of the diamond of 1 TPa. The pressures in excess of TPa and extreme

Femtosecond Laser-Matter Interactions: Solid-Plasma-Solid Transformations at the Extreme Energy Density
Eugene Gamaly

ISBN 978-981-4877-40-4 (Hardcover), 978-1-003-25661-8 (eBook)
www.jennystanford.com

but transient high-pressure/-temperature conditions have been created by the dynamic methods using explosions and megajoule-class lasers (Avrorin et al., 2006; Fortov, 2005; Trunin, 1994; Ahrens, 1980; Salleo et al., 2003; Swift et al., 2012; Hicks et al., 2003). The diamond anvil press and the installations exploring explosions and high-energy lasers are cumbersome and expensive.

It was demonstrated 2 decades ago (Glezer and Mazur, 1997) that the single ultra-short laser pulse tightly focused deep inside a transparent material (depth of the focal spot is much larger than the focal size) creates a nearly spherical void. Later on, the experiments at the laboratory tabletop and the following analysis revealed that the void size in sapphire is the experimental evidence of the pressure in excess of few TPa, well over the Young modulus of sapphire (Juodkazis et al., 2006). The conversion of a transparent dielectric, sapphire, into plasma dramatically increases the laser absorption and the absorbed energy density. Tight focusing of ~200 fs laser pulse with 100 nJ energy deep inside a crystal generates the energy density of several MJ cm^{-3} inside a submicron volume, creating pressure of several TPa. The Young modulus of diamond, the strongest natural material, is 1 TPa. Thus, tight focusing of low-energy femtosecond laser pulse generates pressure several times higher than the strength of any existing material. The confined solid density plasma explodes from the energy deposition volume in a nearly spherical mode, generates a powerful shock wave, and compresses the surrounding pristine material while the rarefaction wave creates the void at the area of the high energy concentration.

There are numerous studies of multiple pulses interactions with the transparent dielectrics at low and high intensities. These studies are out of considerations here. We concentrate in this chapter on the single-pulse interactions, which allow appropriate diagnostics and unambiguous interpretation.

The studies of confined micro-explosion during the past two decades revealed the major features of this complicated phenomenon where the processes of the light focusing, electromagnetic field/dielectric interaction, plasma formation and high-pressure hydrodynamics are intertwined. The schematic sequence of these processes is as follows (Fig. 6.1):

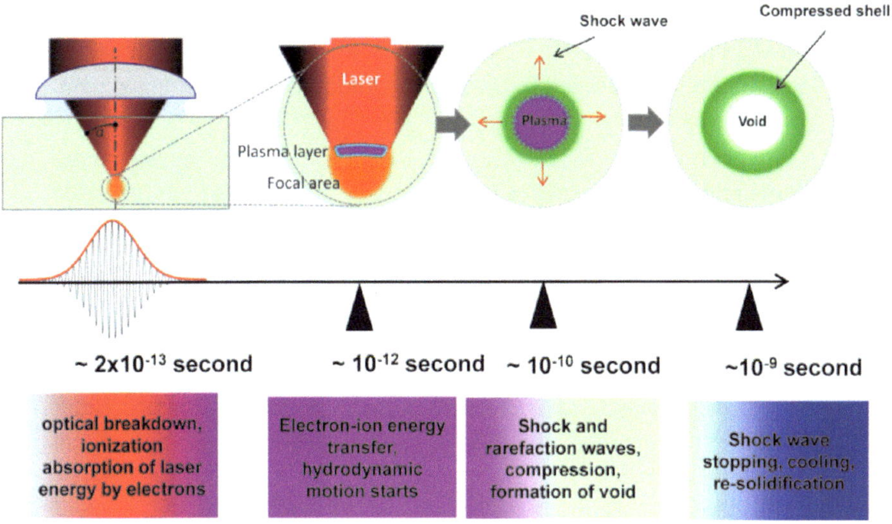

Figure 6.1 Succession and time scales for the processes during confined micro-explosion.

1. The tight focusing of the laser beam several microns deep inside a transparent crystal converts a material to plasma and creates to the end of the pulse the absorbed energy density of few MJ/cm^3 in a sub-micron, nearly spherical volume surrounded by the pristine solid. The absorbed energy during the pulse is confined exclusively in the electrons.
2. During a few picoseconds, the hot electrons transfer the energy to ions.
3. The high pressure of ions and electrons drives the strong shock accompanied by the rarefaction wave.
4. The shock decelerates and converts to the sound wave, when the pressure behind the shock front decreases to the pressure in the cold material. The void, surrounded by the shell of compressed and pressure modified material is formed. All transformed material remains confined inside the bulk of undamaged material ready for the further studies.

These studies initially employed the short intense laser beam with the Gaussian spatial and temporal intensity profile focused inside of a transparent material by the high-NA optics. In Section 6.2.1, we describe interaction of the tightly focused Gauss beam (GB) with the different transparent materials.

Afterwards, the studies were extended on the focusing the GB on the surface of opaque material covered by a transparent layer. Section 6.2.2 is devoted to the interaction of the GB with the silicon layer buried under SiO_2 cover.

Recently, it was shown that the Bessel-shaped pulse could transform much larger amount of a material and supposedly create even higher energy density than that was achieved with the Gaussian (GB) pulses (Courvoisier et al., 2012; Rapp et al., 2016; Gamaly et al., 2017). In Section 6.2.3, we consider the recent development of studies of the Bessel beam (BB) interactions with the transparent materials.

In Sections 6.2.1–6.2.3, we consider the interaction and relaxation processes not discussing the processes of the phase transformations. We focus particularly on the unusual phase transformation path that is unique for the confined micro-explosion: solid-plasma-solid transformation leaving the final product confined inside a pristine crystal.

In Section 6.2.4, we briefly describe the results of the phase state studies of the materials transformed by the confined micro-explosions. The studies of the laser-affected materials, which led to discoveries of novel phases, have been performed by x-ray and electron diffractions well after the interaction experiments. The novel body-centred-cubic (bcc) aluminium, and two previously unobserved phases of silicon were discovered.

We conclude discussing the future directions.

6.2 Interaction of the Gauss Beam Tightly Focused inside a Transparent Crystal

In what follows, by the GB, we refer to the beam having the Gauss-like intensity distribution in the space across the incident beam and in time. In order to produce a micro-explosion inside a transparent crystal the energy of a laser beam should be first delivered inside the transparent crystal and concentrated inside in a diffraction limited focal spot at the depth 5–50 μm. In what follows in this section, we describe the succession of the processes resulting in the high energy density in the energy absorption volume. We consider first the delivery of the pulse to the focal spot. Then the intensity distribution in the focal volume created by the high-NA optics shall be defined. The tight focusing creates at the focal surface the energy density well in excess the ionization threshold. Therefore, the permittivity of the laser-affected dielectric is modified being converted to plasma. The running electromagnetic wave transforms to the evanescent wave and the major absorption occurs.

6.2.1 Beam Propagation

The intensity in tightly focused beam significantly increases during the propagation. In order to avoid the energy losses during the delivery of the pump laser energy to the focal spot, the power of the laser should be kept lower than the self-focusing threshold for the medium. The critical value for the laser beam power signifying the onset of the self-focusing depends on the non-linear part of the refractive index, n_2 ($n_2 = n_0 + n_2 I$), and on the laser wavelength, λ, as follows (Akhmanov et al., 1988):

$$p_{\mathrm{sf,cr}} = \frac{\lambda^2}{2\pi n_2 n_0} \tag{6.1}$$

The self-focusing begins to affect the medium when the power of the laser beam exceeds the critical value, $p > p_{\mathrm{cr}}$. If the power of the Gaussian beam exceeds the critical power, then the beam self-focuses after propagating the distance (Akhmanov et al., 1988):

$$L_{\mathrm{sf}} = \frac{2\pi n_0 r_{\mathrm{min}}^2}{\lambda}\left(\frac{p}{p_{\mathrm{sf,cr}}} - 1\right)^{-1/2} \tag{6.2}$$

For example, in fused silica (n_0 = 1.75, n_2 = 3.54 × 10^{-16} $\mathrm{cm^2/W}$ for λ = 1 μm) the critical power comprises 3.18 MW. Taking r_{min} = 1 μm, p = 2 × $p_{\mathrm{sf,cr}}$, one gets L_{sf} = 9 μm equal approximately to the depth of the in-bulk focusing with the GB. In the experiments with the GB (200 nJ, 200 fs), the beam power was well below the critical value (Juodkazis et al., 2006).

6.2.2 Focusing of the Gauss Beam with the High-NA Optics

6.2.2.1 Conventional approach to the high-NA focusing

Single laser pulse can be tightly focused inside a transparent crystal using the optical microscope equipped with the oil-immersion objective lens of the high numerical aperture. The numerical aperture by definition is NA = $n_0 \times \sin\alpha$. Here, n_0 is the refractive index in the unperturbed crystal, and α is a half of the convergence angle.

The refractive index mismatch between the immersion oil (n = 1.52) and sapphire (n = 1.75) causes a strong dependence of the focal intensity distribution on the depth of irradiation due to spherical aberrations. The deep focusing led to the elongated voids. The smallest possible length of 5–10 μm was used in the experiments allowing minimizing the chromatic and, mainly, spherical aberrations of the focusing optics. The scheme of the device using optical microscope and oil-immersion objective lens of the high numerical aperture is presented in Fig. 6.2 (Juodkazis et al., 2006).

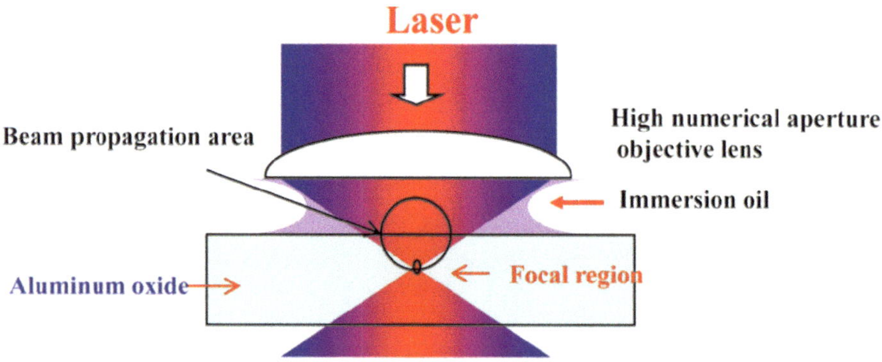

Figure 6.2 Focusing the laser beam inside a transparent dielectric with the help of the optical microscope and the high-numerical-aperture optics.

A plane wave focused by a lens creates at the focus the axially symmetric intensity distribution (Born and Wolf, 2003). The intensity distribution across the focal spot is the same as that for the diffraction of a plane wave on a round aperture. Therefore, it is reasonable to take the radius of the Airy disk (the first minimum of the point spread function), as a beam waist radius. The plane wave can be focused to the diffraction limit, which depends on the focusing angle (through NA defined above). Then the focal spot radius (Airy disk radius) is as follows:

$$r_{\mathrm{foc}} = \frac{0.61 \cdot \lambda}{\mathrm{NA}} \tag{6.3}$$

We define the focal volume in a material with the optical properties unaffected by laser as a space confined inside the surface where the intensity equals a half of its maximum value (see Fig. 6.3).

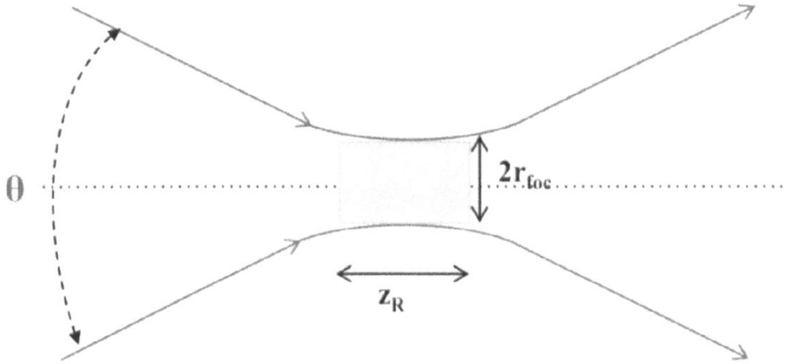

Figure 6.3 Scheme of the focal area in a laser-unaffected medium. θ is the convergence angle.

Within the Airy disk, the 83% of the energy of the incident plane wave is enclosed. On the other hand, the laser beam with the Gauss intensity distribution contains 86% of energy within a disk where the intensity equals $1/e^2$ of the maximum. Therefore, it is reasonable to take the above formula as the radius of the GB. The length of the focal volume is conventionally taken equal to the doubled Rayleigh length, presented as follows:

$$Z_R = \frac{\pi r_{\mathrm{foc}}^2 \times n_0}{\lambda_0} \tag{6.4}$$

For sapphire (n_0 = 1.75) at 800 nm one obtains r_f = 0.38 μm; z_R = 1.737 μm. However, at high intensity, a dielectric is converted to plasma and the absorption length appears to be much shorter than the Rayleigh length. The focal spot area allegedly remains close to the unaffected case because the heat conduction during the pulse remains insignificant. The focal spot area for the tightly focused GB is as follows (some geometrical corrections were introduced by Gamaly et al., 2006):

$$S_{\text{foc}} = \frac{0.405 \cdot \lambda_0^2}{\text{NA}^2} \tag{6.5}$$

For example, taking λ_0 = 800 nm, NA = 1.35, one gets the focal spot area of S_{foc} = 0.142 μm^2. Therefore, 100 nJ laser energy focused at such spot delivers the surface energy density (fluence) of F = 70.4 J/cm^2 to the end of the pulse. It means that at the beginning of the pulse the surface energy density well in excess of the ionization threshold ($F_{i,\text{th}}$ = 0.084 J/cm^2) and of that necessary for conversion to plasma (F_{plasma} = 0.39 J/cm^2) has been delivered. The optical properties are drastically modified and the absorption depth decreases to l_{abs} = 31 nm (see details in the following section). Hence, early in the pulse time in accord with above calculations the absorption volume consist of the flat disk of diameter $2r_f$ = 760 nm with thickness of 31 nm. However, these calculations did not take into account the possible modifications of the intensity distribution across the tightly focused beam. Apparently, they are not complying fully with the experimental observation.

6.2.2.2 Experimental evidence: high-NA focus produces nano-sharp intensity distribution

Let us discuss the amazing experimental observations of the voids produced with the high-NA (1.35) focusing, which did not get the attention they deserve. First, the void of ~200 nm in diameter has a form close to a *perfect sphere*. Second, the laser-affected area surrounding the void (containing amorphous sapphire, therefore being affected by rather moderate intensity) has a length approximately equal to the Rayleigh length as in sapphire with unchanged optical properties (see Fig. 6.4a) (adapted from Joudkazis et al., 2006; Gamaly et al., 2006). The observed picture of the crystal modification is an imprint of the final spatial intensity distribution and absorbed energy density, which resulted in the detected structure.

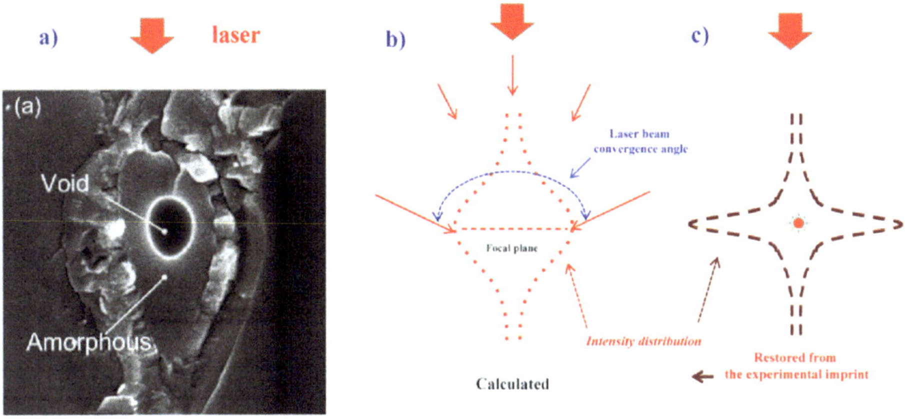

Figure 6.4 Recovery of the intensity distribution produced by the high-NA focusing from the structure imprinted in the sapphire crystal by the energy deposited in the focal region: (a) void surrounded by the amorphous sapphire produced by 100 nJ, 200 fs laser, focused with NA = 1.35 (Juodkazis et al., 2006); (b) intensity distribution calculated in accord with classical optics (schematic) (Juodkazis et al., 2006); (c) intensity distribution recovered from the experimental imprint of the real energy deposition.

First, observation indicates that the high-energy-density area, located at the centre of the void, was significantly smaller than the void diameter (<<200 nm) ensuring the *central-symmetric-like explosion*, which led to the nearly spherical void formation. The focal spot diameter for the plane wave, focused in sapphire with high NA = 1.35 optics, is around 760 nm, calculated in accord with the classical optics. Hence, the measured transverse diameter of the spot, where intensity (and energy deposition) has a maximum, is many times smaller than the diameter of the calculated focal spot (760 nm). However, relatively low-intensity area has transverse and axial dimensions comparable to the focal volume calculated in accord with the classical optics.

The calculated absorption depth (via Landau–Spitzer–Drude formulae) in sapphire converted to solid density plasma is in a range 20–40 nm, and it is much smaller than the measured void size complying with the necessary conditions for the energy deposition area size for producing nearly spherical void. The experiments directly indicate that the transverse size of the maximum intensity area is much smaller than it follows from the existing understanding of the high-NA focusing.

Several theoretical papers demonstrated that high-NA focusing drastically changes the internal structure of the beam. The authors of the reference (Bomzon et al., 2006) have shown that the energy distribution in the focal plane of a beam, focused with high-NA lens possesses a form depending on polarization. Circularly polarized light focal spot has a circular symmetry while the linearly polarized light produces the rectangular-like distribution at NA = 0.95. Focusing a beam through a high-NA lens also causes space-varying polarization and depolarization. The same authors found the threshold when the increase in NA starts changing the amount of the geometrical phase in the beam. The ratio between the portion of the beam possessing of the angular momentum and energy of the entire beam might be considered as a fraction of the beam possessing the Berry phase. They found that this ratio monotonically increases until a value of about 1.4 at NA = 0.95. The authors explained this apparent paradox showing that the lens acts as a low pass filter. The lens only transmits spatial frequencies lower than cut-off wave number. The high spatial frequencies carry more angular momentum, related to the geometrical phase, per unit energy than the low spatial frequencies. Cut-off wave number increases with NA. Therefore, increasing the NA will inevitably increase the amount of the orbital angular momentum (OAM) per unit energy present in the focal region.

Nieminen et al. (2008) demonstrated that the strongly focused circularly polarized Gaussian beam with a half of the convergence angle of 45° (NA = 1.237 in sapphire) converts into the longitudinal vortex possessing the topological phase, phase singularity and the axial component of the electric field. One might conjecture that at the very large convergence angles (NA ≈ 1.4) the intensity distribution across a focused beam changes from the Gauss distribution to the Bessel distribution (Bessel function of the first kind zero order) with the maximum in the centre of the void.

Therefore, the transformation of a transparent crystal under the action of the beam focused by the high-NA optics takes place on the intensity landscape drastically different from the current understanding. Conversion to plasma first occurs near the intensity maximum and the following essential energy absorption happens in the solid plasma with the increasing electrons number density. There are no studies helping determining the size of such a sharp focus either theoretically or experimentally.

Many research groups tried to find the analytical and numerical solutions for the Maxwell equations describing the tight focusing beyond the paraxial approximation (see Gonzalez et al., 2018 and references therein). The solutions to the Maxwell equations presented by Gonzalez et al. (2018) conserve the energy and symmetry and significantly improve the paraxial approximation. However, the numerical example of the tight focusing presented (800 nm, 20 fs, 36 nJ laser beam focused at NA = 0.57) does not show the effects mentioned above, which become apparent at NA > 0.95.

On the current level of understanding, only Heisenberg's uncertainty principle allows estimating the limit for the minimum space scale achievable by the photons concentration under peculiar high-NA focusing: $\Delta x \geq \lambda/4\pi$, and therefore the minimum size of the sharp focus. The minimum size of the laser-affected zone in sapphire (n = 1.75) at 800 nm constitutes $\Delta x = \lambda/4\pi \times n$ = 36.38 nm. It is reasonable to suggest that the size of the energy deposition area in sapphire under high-NA focusing is rather close to this limit. Note that the electric field distribution in the plasma area (where the real part of the dielectric permittivity is negative) is governed by the modified Bessel function of the first kind zero order (Morse and Feshbach, 1953). Thus, the above assumptions suggest the energy deposition volume to the end of the pulse being a cylinder with diameter $\Delta x = \lambda/4\pi \times n$ = 36.38 nm and the absorption length $l_{abs} \approx 30$ nm. The whole absorbed energy to the end of the pulse is confined in the electron sub-system. The electronic heat conduction in created plasma starts to be efficient during the pulse and after the end of the pulse. It leads to the smoothing out the electron temperature and the energy density over the absorption volume.

The hydrodynamic motion, leading to the void formation, begins after the energy transfer from the hot electrons to cold ions. The electron-to-ion energy transfer time lasts a few picoseconds. Major smoothing out the electron temperature over the absorption volume occurs during the energy transfer time and electron/ion temperature equilibration. Taking conservatively heat diffusion coefficient in plasma $D \approx 3\ \text{cm}^2/\text{s}$ one gets the heat diffusion length during 3 ps equal to $l_{\text{therm}} \approx (D \times t)^{1/2} \approx 30$ nm. It means that the heat diffusion homogenizes the absorption volume making it close to the sphere of 30–40 nm diameter, the perfect spherical energy source for creating nearly spherical void. The processes of the shock and rarefaction wave generation and formation of void then follow the scenario describing the void formation in (Juodkazis et al., PRL, 2006).

To conclude the modification of the laser beam structure via the high-NA focusing has been imprinted into laser-modified material. The measured *post mortem* void size and large laser-affected area allow restoring qualitatively the intensity (absorbed energy density) distribution (see Fig. 6.4). It is reasonable to expect that the high-NA focusing at low intensity produces similar sharpening of the intensity distribution in a focal volume resulting in the nano-size laser-affected zone. Theoretical studies and further experiments of the focusing at the large convergence angles are needed to improve the existing understanding.

6.2.3 Laser Modification of the Dielectric Permittivity: Conversion to Solid Density Plasma

Wide-band gap dielectrics are transparent for the visible light at low intensity because the energy of photons is much less than the band gap. Therefore, the permittivity of such dielectrics at a low intensity possesses a large positive real part and negligibly small imaginary part. The only way for increasing the absorption is increasing the incident intensity to a level, when the sufficient number of the valence electrons absorbs the multiple photons, their energy exceeds the band gap and the electrons are transferred from the valence to the conduction band. Then the incident light begins absorbing on the conductivity electrons due to the inverse bremsstrahlung process. Under such conditions, the properties of the material and the laser–material interaction change rapidly during the pulse.

The multi-photon ionization (MPI) and ionization by the electron impact (IEI) are the major processes responsible for the electrons transfer from the valence to the conduction band (see Chapter 3 for more details). MPI has no threshold and hence the contribution of multi-photon ionization can be important even at relatively low intensity. Multi-photon ionization creates the initial (seed) electron density, which then grows up by the impact (avalanche) process. Continuous increase in the intensity from the beginning of the pulse modifies the permittivity of a dielectric gradually reducing the positive polarization contribution of the valence electrons and the increasing the negative contribution of the conduction electrons. The real part of the modified permittivity turns to zero when growing negative electronic part balances the positive polarization part.

6.2.3.1 Threshold for Conversion of Dielectric into Absorbing Medium ($\varepsilon_{re} = 0$)

At the moment, when $\varepsilon_{re} = 0$, the laser wave converts from the running wave to the evanescent wave. The intensive absorption begins at the surface separating dielectric and plasma at $\varepsilon_{re} = 0$. The event, when the condition $\varepsilon_{re} = 0$ is fulfilled, is defined as the ionization threshold. This condition allows determining the number density of electrons transferred from the valence band to the conduction band. The approximate formula for the permittivity from Chapter 3 gets the number density of conduction electrons at this state, $n_e \approx \varepsilon_0 \times n_{cr}$, expressed through the dielectric permittivity of the unexcited dielectric and through the critical electron density corresponding to the frequency of the incident light. For example, in sapphire ($\varepsilon_0 = 3.0625$; $n = 1.75$) under the action of 800 nm light (critical density, $n_{cr} = 1.75 \times 10^{21}$ cm^{-3}) the number density of conduction electrons in this state constitutes, $n_e = 5.36 \times 10^{21}$ cm^{-3}, that is ~4.47% from the atomic density. Zero real part of the permittivity results in the equality of the real and imaginary parts of the refractive index, $n = \kappa = (\varepsilon_{im}/2)^{1/2}$. The imaginary part of the permittivity at $\varepsilon_{re} = 0$ is as follows:

$$\varepsilon_{\text{im}}(\varepsilon_{\text{re}} = 0) = \frac{\varepsilon_0 \times \nu_{\text{eff}}}{\omega \times (1 + \nu_{\text{eff}}^2 / \omega^2)}$$

The conventional kinetic relation, $\nu_{\text{eff}} \approx n_e \cdot \sigma \cdot v_e$, allows determining the electrons collision rate in these conditions. Taking electron density as above, collision cross section as for the atomic collisions, $\sigma \approx 10^{-15}$ cm^2, and velocity of electron in the conduction band as $v_e \approx 2 \times 10^8$ cm/s, one obtains $\nu_{\text{eff}} \approx 1.07 \times 10^{15}$ s^{-1}. Thus, the imaginary part is $\varepsilon_{\text{im}} = 1.15$, and $k = 0.76$ giving the absorption depth in plasma, $l_{\text{abs}} = c/\omega \cdot \kappa = 167.6$ nm and absorption coefficient, $A = 0.844$.

The energy density necessary for transferring, $n_e \approx \varepsilon_0 \times n_{\text{cr}}$, electrons through the energy band gap of Δ_{g} then equals $\varepsilon_0 \times n_{\text{cr}} \times \Delta_{\text{g}}$ [J/cm^3]. On the other hand, absorbed energy density is $E_{\text{abs}} = 2A \cdot F/l_s$ (see Chapter 2). Equalizing both formulae, one can find the value of the threshold laser fluence necessary for the conversion of dielectric into absorbing metal-like medium, $F_{i,\text{thr}} = \varepsilon_0 \cdot n_{\text{cr}} \cdot \Delta_{\text{g}} \cdot l_s / 2A$. Taking parameters for sapphire ($\varepsilon_0 = 3.0625$; $A = 0.844$; $l_{\text{abs}} = 167.6$ nm; $\Delta_g = 9.9$ eV; $\varepsilon_b \approx 4$ eV; $J_i \approx 11.6$ eV), one gets $F_{i,\text{th}} = 0.084$ J/cm^2. Note that the absorbed energy density comprises 8.5 kJ/cm^3.

The further increase in the intensity during the pulse results in transforming a dielectric into the solid density plasma by moving all valence electrons first into the conduction band and after that into continuum. The additional energy exceeding the cohesive energy per atom (around 3–5 eV/atom) should be delivered in order to break the atomic bonds.

6.2.3.2 Conversion a dielectric to the solid density plasma (single electron stripped off any atom)

In order to transform a dielectric to solid density plasma it is necessary to deliver to each atom the energy at least equal to the first ionization potential plus the cohesive energy, $\varepsilon_b + J_{1,i}$, or energy density, $n_a \times (\varepsilon_b + J_{1,i})$. For example, the energy density, necessary for the complete single ionization of sapphire (Al_2O_3), equals ~ 0.17 MJ/cm^3. Note that along with the increasing free electrons number density the other plasma parameters, collision rate and dielectric permittivity are also changing. Parameters of the sapphire plasma at the state of complete single ionization of all atoms (one electron is stripped off each atom) are the following: $n_e = 10^{23}$ cm^{-3}; $\nu_{\text{eff}} = 5 \times 10^{15}$ s^{-1}; $\omega_{\text{pe}} = 1.78 \times 10^{16}$ s^{-1}. Optical characteristics of such a plasma are: $\varepsilon_{\text{im}} = 22.07$; $\varepsilon_{\text{re}} = -9.4$; $n = 2.27$; $k = 4.086$; absorption length comprises $l_{\text{abs}} = 31$ nm and absorption coefficient, $A = 0.33$. The threshold fluence for the full conversion sapphire to plasma follows from the condition $2A \cdot F_{p,\text{th}}/l_s \approx n_a \cdot (\varepsilon_{\text{b}} + J_{1.i})$ and comprises, $F_{p,\text{th}} = n_a \cdot (\varepsilon_{\text{b}} + J_{1.i})\, l_s/2A = 0.39$ J/cm^2.

6.2.3.3 Ionization after the end of the pulse

The electron's temperature at the end of the pulse is much higher than the ionization potential; therefore, the ionization by the electron impact continues after the end of the pulse. The rate equation for the electron number density accounting for the impact ionization and for the recombination to the atomic ground state is as follows (Zel'dovich and Raizer, 2002):

$$\frac{dn_e}{dt} \approx \alpha_e n_e n_a - \beta_e n_i n_e^2 \tag{6.6}$$

Here, $\alpha_e = \sigma_e v_a \left(\frac{J_z}{\varepsilon_e} + 2\right) \times \exp\left(-\frac{J_z}{\varepsilon_e}\right)[\text{cm}^3/\text{s}]$ is the impact ionization rate and β_e is the recombination coefficient. The ionization and recombination coefficients, α_e and β_e, are related by the principle of the detailed balancing, $\beta_e = \alpha_e/K(\varepsilon_e)[\text{cm}^6/\text{s}]$, $K(\varepsilon_e)[\text{cm}^{-3}]$ is the equilibrium constant following from the Saha equation (Zel'dovich and Raizer, 2002). For our analysis, it is important that the avalanche development time, $\tau_{\text{av}} \approx (\alpha_e n_e)^{-1}$, also characterizes the time for the establishment of the ionization equilibrium. In the considered case, the electron energy is several times larger than the ionization potential. Therefore, the cross section for the impact ionization is close to its maximum of $\sigma_e \sim 2 \times 10^{-16}$ cm^2. One can see that for the plasma parameters in question the avalanche time as well as the ionization equilibration time is very short $\tau_{\text{av}} \approx \tau_{\text{eq}} \leq 10^{-15} \cdot s$. Hence, an average ion charge can be estimated using the Saha equations where the ionization balanced by the recombination.

The plasma state after the end of the pulse characterizes by the electrons number density, average ion's charge, $z = n_e/n_i$, and the electrons average energy. Number density of ions remains constant before beginning of the hydrodynamic motion. Set of two coupled equations, the Saha equation and energy balance equation accounting for the ionization losses, allows determining two unknowns, ion's charge and electron energy. These cumbersome calculations are simplified by the approach, in which the discrete potentials of mono-atomic or complex solid are interpolated by the continuous function of the average ion charge, $J_k(z + 1/2)$. Subscript k denotes the kind of ion (Zel'dovich and Raizer, 2002). The modified Saha equation expresses the average ion charge through the electron energy (in eV):

$$J_k(z + 1/2) = k_B T_e \ln \frac{A(k_B T_e)^{3/2}}{z n_i} \tag{6.7}$$

The constant is, $A = 6 \times 10^{21}$ cm^{-3}eV$^{-3/2}$. These calculations suggest that after the end of the pulse additional 3–4 electrons are stripped off each atom (Gamaly et al., 2006). We are considering the plasma state shortly after the end of the pulse while the processes of the heat conduction and electron-to-ion energy transfer are still slow (characteristic times for these processes are in a few ps range). The electrons' internal energy equals the absorbed energy minus the ionization losses.

$$C_e \cdot n_e \cdot k_B T_e \approx E_{\text{abs}} - E_{\text{ion,loss}} \tag{6.8}$$

Ionization losses for silica and sapphire (assuming stripping 3–4 electrons from each atom) are in a range of (0.5–0.6) MJ/cm^3 that is around a few percents of the absorbed energy density.

6.2.4 Relaxation Processes

The state of the laser-affected sapphire crystal to the end of the pulse is the solid density plasma with 3–4 electrons, stripped off the each atom. Part of the absorbed energy has been spent on this transformation (~0.5 MJ/cm^3). Now several relaxation processes are going simultaneously. These processes are the energy transfer from electrons to ions, energy transport by electrons and electrons recombination with ions. Our goal in this paragraph is defining the plasma state at the start of the ions' motion finally resulting in the formation of the observed structure. The absorbed energy is recovered from the experimentally observed structure, that is the void surrounded by the shell of the compressed amorphous sapphire. This structure is the evidence that the absorbed energy, concentrated in a small volume after focusing and absorption, then generated a strong shock. The shock wave moved in the surrounding material, compressed it, and gradually decelerating finally stopped when the pressure behind the shock front turned equal to the cold pressure in a crystal. The surface, where this balance was achieved, allegedly coincides with the boundary of the laser-affected area experimentally observed as the outer boundary of the amorphous sapphire. Therefore, the energy confined in the spherical volume inside this surface is estimated conservatively (minimizing) as a cold pressure (the Young modulus of material, Y) by this volume, $E_{\text{abs}} \geq \frac{4\pi}{3} r_{\text{stop}}^3 \cdot Y$, that is comparable to the minimum absorbed energy. Therefore, the void and shell formation is completely described by the absorbed energy, density and cold pressure of material.

100 nJ pulse produced a void in sapphire surrounded by a shell with the thickness equal to the void radius evidencing the absorption of at least 50% of the laser energy (Juodkazis et al., 2006). However, the volume where this energy was initially (before the ions' motion) contained to the end of the pulse is unknown as well as the maximum energy density (pressure) and the electron temperature. Knowledge of the pressure/temperature range allows define conditions for the material transformations at extreme. The density remains equal to the initial solid density because the ions are still cold. Absorbed energy density rate defined for the plane wave absorption above can serve for the estimation of the range for the maximum temperature/pressure in the absorption volume.

6.2.4.1 Range for the maximum temperature/pressure value in the absorption volume

The absorbed energy density rate is as follows: $\frac{dE_{\text{abs}}}{dt} = 2A \cdot I(t)/l_{\text{abs}}$ (see Chapter 2). Integrating over the pulse duration and taking the absorption coefficient and length as the average over the pulse time, one gets the absorption energy density in the convenient form:

$$E_{\text{abs}} = 2A \cdot E_{\text{las}} / S_{\text{foc}} \cdot l_{\text{abs}} = 2A \cdot E_{\text{las}} / V_{\text{abs}} \tag{6.9}$$

All energy is in the electrons. Therefore, electrons' temperature is estimated as follows:

$$k_B T_e \approx A \cdot E_{\text{las}} / C_e \cdot n_e \cdot V_{\text{abs}} \tag{6.10}$$

The electrons' number density to the end of the pulse is in a range $n_e = 10^{23}$ to 4×10^{23} cm^{-3} depending on the ionization degree. Electron heat capacity is close to that of the ideal gas, $C_e \approx 3/2$. If one takes the absorption volume equal to the estimated above focal volume at the diffraction limit of $V_{\text{abs}} \approx 10^{-15}$ cm^{-3} the electron energy in a range of $E_e \approx 10^4$ eV immediately follows. However, the electrons mean free path for the Coulomb collisions grows as the square of the electron energy, $l_e \sim E_e^2$. The electron energy swiftly increases after the conversion to plasma (at ~10 eV, electron's mean free path, $l_e \approx 1$ nm) spreading the energy on the larger volume. Electrons with the energy around 100 eV have the mean free path around 100 nm. The energy transfer by electrons becomes the dominating mechanism defining the energy deposition volume. Thus, it is reasonable to estimate the size of the volume taking the sum of the focal spot diameter and electron mean free path to the end of the pulse as a diameter of the spherical volume $V_{\text{abs}} \approx 10^{-15}$ cm^3. Hence, half of 100 nJ laser pulse energy (absorption coefficient, $A = 0.5$) creates the electrons' energy density of 50 MJ/cm^3 (50 TPa of pressure). Assuming that three electrons are stripped from aluminium and oxygen atoms to the end of the pulse (ionization losses constitute around MJ/cm^3) the electrons energy reaches few hundreds of eV. Note that the above estimates are semi-quantitative. Therefore, it is realistic assuming the electron pressure and temperature is within the ranges, P_e = (5–50) TPa; $k_B T_e \approx (10^2–10^3)$ eV while the ions are cold. Three intertwined processes are responsible for the energy transfer from the electrons to ions: action of the electrostatic force of the electronic pressure gradient; the Coulomb collisions transferring the momentum and transporting the energy by the means of the heat conduction and the Coulomb collisions transferring the energy (Gamaly et al., 2006).

6.2.4.2 Ion's acceleration by the gradient of the electronic pressure

We estimate the energy transfer from the energetic electrons to cold ions by the action of the electronic pressure under suggestion that hydrodynamic approximation is legitimate in these dynamic conditions. The Newton equation for ions with mass velocity and number density respectively, M_i, u_i, n_i, reads

$$\frac{\partial M_i n_i u_i}{\partial t} \approx -\nabla P_e.$$

We estimate $\nabla P_e \approx n_e k_B T_e / l_{\text{abs}}$. During the time period of $t_{\text{el-st}}$, the velocity of ion reaches the value:

$$u_i \approx \frac{Z k_B T_e t_{\text{el-st}}}{M_i l_{\text{abs}}}$$

After the temperature equilibration between $(Z + 1)n_i$ particles, the common temperature equals $T \approx ZT_e/(Z + 1)$. The ion velocity at that moment $u_i(t_{\text{el-st}}) \approx \left(\frac{2k_B T}{M_i}\right)^{1/2}$ allows the determination of the time when the equilibration is achieved:

$$t_{\text{el-st}}(T) \approx l_{\text{abs}} \left(\frac{2M_i}{k_B T_e}\right)^{1/2}$$

One gets that for the temperature range, $k_B T_e \approx (10^2–10^3)$ eV, the energy transfer time to ions by the electronic pressure is in a few picoseconds range, $t_{\text{el-st}} = 2–20$ ps.

6.2.4.3 Electrons-to-ions energy transfer by the Coulomb collisions: Spatial separation of the light and heavy ions

First, we shall note that the dense high-temperature plasma in the above conditions is in the ideal plasma state. Indeed, plasma is in ideal state when the number of particles in the Debye sphere is large, $N_D >> 1$. Calculating this number at $n_e = 3 \times 10^{23}$ cm^{-3}, $k_BT_e = (10^2–10^3)$ eV, one gets $N_D = 1.7 \cdot 10^9\ [(k_BT_e)^3/n_e]^{1/2} \approx (3–10^2)$. Therefore, the Coulomb collision rate is taken as for the ideal plasma in the following form (Kruer, 1988):

$$\nu_{\text{ei}} \approx 3\times10^{-6}\cdot \text{In}\Lambda \frac{z\cdot n_e}{(k_BT_e)^{3/2}_{\text{eV}}}\left[s^{-1}\right]$$

At $z = 3$; $n_e = 3 \times 1.18 \times 10^{23}$ cm^{-3}; $k_BT_e = 10^3$ eV one gets the collision rate, $\nu_{\text{ei}} = 2 \times 10^{14}$ s^{-1}. The electron-to-ion energy exchange rate in plasma is expressed via the electron–ion momentum exchange rate as follows (Landau, 1937):

$$\nu_{\text{ei}}^{\text{en}} \approx \frac{m_e}{m_e}\nu_{\text{ei}}$$

Therefore, in plasma with the mixture of different ions the light ions get energy from the hot electrons and establish equilibrium faster than heavy ions. In sapphire (Al_2O_3, at $n_e = 3 \times 10^{23}$ cm^{-3}, $k_BT_e = 10^2–10^3$ eV), the electron–oxygen ion equilibrating time is, $t_{\text{e-O}} = (\nu_{\text{e–O}}^{\text{en}})^{-1} = 5.28–167$ ps, while for the electron–aluminium ion the equilibration time equals $t_{e\text{-Al}} = 7.9–250$ ps. Light ions start moving and diffusing earlier resulting in the spatial separation of light and heavy ions.

The processes of the energy transfer are going simultaneously and affecting each other. Note that the time of the energy transfer by the electrostatic force of the electronic pressure and by the Coulomb collisions depends on the electron temperature differently. Indeed, the electrostatic time is inversely proportional to the electron temperature, $t_{\text{el-st}} \propto M_iT_e^{-1/2}$, while the collision time is $t_{\text{e-i}} \propto M_iT_e^{3/2}$.

Therefore, the effective time of the energy transfer to ions should be considered as an average of all times.

6.2.4.4 Diffusion of ions with the different masses: The experimental evidence of the spatial separation

The formation of the tens of nanometer-sized crystallites of super dense aluminium out of aluminium oxide observed in the experiments (Valionis et al., 2011) is the evidence of the spatial separation of oxygen and aluminium ions in conditions of complete confinement and preserved stoichiometry. The primary mechanism of light and heavy ions spatial separation relates to the Coulomb collisions responsible for the energy transfer from electrons to ions. The other process leading to the further spatial separation of light and heavy ions is diffusion. The light oxygen ions after getting the energy from keV electrons pick up velocity of v_O = 73 km/s and continue diffusing. The mean free path for the ions collision in the solid density plasma is the inter-atomic distance, d_a. Therefore, the diffusion coefficient takes a conventional form of, $D_O = d_a v_O/3$. Taking $d_a \approx 2.15$ Å, one obtains $D_O = 0.052$ cm^2/s. Subscript *"O"* stands for oxygen and Al for aluminium. During 83 ps oxygen ion diffuses to the distance $L_{diff} = \{D_0 \times (t_{Al} - t_O)\}^{1/2} \approx 21$ nm well before aluminium ions start moving. Therefore, aluminium and oxygen atoms became spatially separated making formation of the tens-of-nm sized crystallites of super-dense aluminium viable. Note that after the completion of the energy transfer, aluminium ions have the diffusion coefficient 1.3 times lower than that for oxygen. Thus, the distance of the ions separation may only slowly increase. The pressure during this period remains in the TPa range being appropriate for the dense phase formation. The hydrodynamics, joint ionic motion, shock wave, starts after the aluminium ions receive the energy from electrons. The temperature of electrons and all ions equilibrates, and 250 ps after the end of the pulse, the shock wave is generated. However, the structure of the shock-wave-front on the nanoscale is complicated: The electrons are moving first, followed by the oxygen ions and then trailing by the aluminium ions schematically presented in Fig. 6.5.

These considerations show that the electron–ion energy equilibration occurs in a volume with a radius approximately 120 nm (energy density 13.8 MJ/cm^3) and initial shock velocity being around 60 km/s. After electron–ion equilibration plasma attains thermal equilibrium state and further motion can be described in the frame of the macroscopic hydrodynamics.

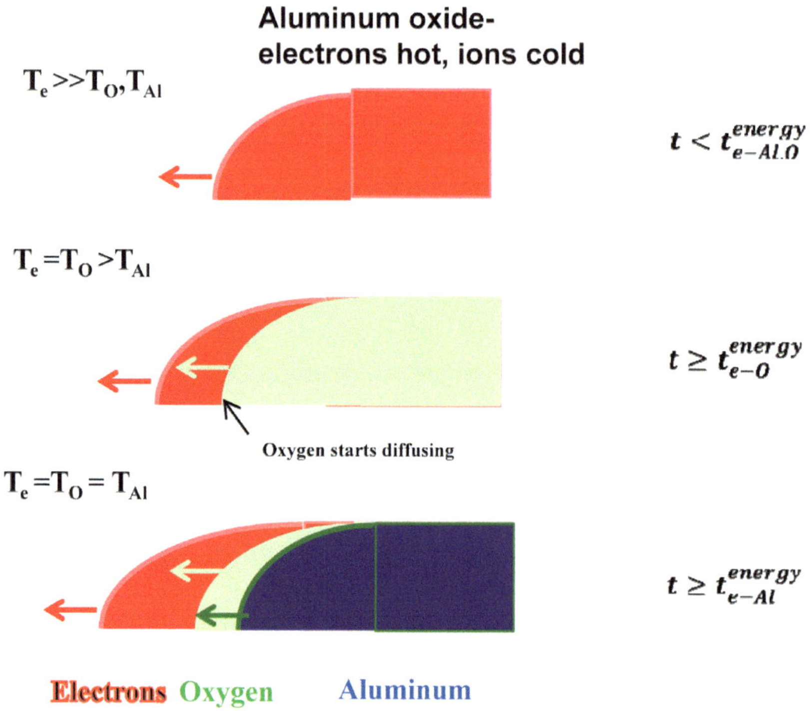

Figure 6.5 Formation of the shock wave from the non-equilibrium state, when the whole absorbed energy is in electrons while the oxygen and aluminium ions are cold. During the equilibration process via the Coulomb collisions, the spatial separation of light and heavy ions occurs resulting in the complicated structure of the shock wave (red, electrons; light green, oxygen; dark blue, aluminium).

6.2.5 Micro-Explosion in Sapphire Considered as a Strong Point-Like Explosion

The structure imprinted by the laser in the sapphire crystal (see Fig. 6.6.) bears a strong resemblance to the frozen picture of the strong point explosion as it emerges from the analytic and numerical solutions in the frame of macroscopic hydrodynamics (Sedov, 1959; Zel'dovich and Raizer, 2002; Landau and Lifshitz, 1980). The experiments and theory demonstrated that the high energy deposited in a volume much smaller than that embraced by the expanding shock wave can be semi-quantitatively described by the analytic and numerical solutions in the frame of the strong point explosion problem.

The exact formulation of the point explosion problem is as follows. The energy source with the finite energy, E, locates at the point in the infinite perfect gas, characterized by the ratio of the specific heat at constant pressure and constant volume, $\gamma = c_p/c_V$, and the mass density, ρ. The energy is rapidly released at the centre of the sphere. The pressure behind the shock is much larger than the initial pressure, p_0, in a medium. Velocities of both shock front and material behind the shock are much larger than the initial speed of sound in a medium, c_0. The length scale is absent making the problem depending only on the ratio E/ρ and γ in the form of the dimensionless similarity variable, ξ:

$$\xi = r\left(\frac{\rho_0}{E \cdot t^2}\right)^{1/5} \tag{6.11}$$

The solution depends only on the single dimensionless parameter ξ_0. This parameter is determined from the condition of the energy conservation and depends only on γ.

The shock front position determined as follows:

$$R = \xi_0(\gamma) \cdot \left(\frac{E}{\rho_0}\right)^{1/5} \cdot t^{2/5} \tag{6.12}$$

The shock front velocity is

$$D = \frac{dR}{dt} = \frac{2}{5}\frac{R}{t} = \frac{2}{5}\xi_0\left(\frac{E}{\rho_0}\right)^{1/5} \cdot t^{-3/5} = \frac{2}{5}\xi_0^{5/2}\left(\frac{E}{\rho_0 \cdot R^3}\right)^{1/2}. \tag{6.13}$$

Pressure, density and velocity behind the shock are taken in the strong shock approximation neglecting the initial pressure in a medium ($p >> p_0$):

$$\rho_1 = \rho_0 \frac{\gamma+1}{\gamma-1};\ p_1 = \frac{2}{\gamma+1}\rho_0 D^2;\ u_1 = \frac{2}{\gamma+1} \cdot D \tag{6.14}$$

The solution determines space dependence of density, pressure and velocity behind the shock front.

The density behind the shock front decreases fast to zero at the central point. Practically entire mass short after the explosion is contained within a thin spherical shell near the shock front, similar to the structure produced by the confined micro-explosion. In the exact solution, there is a range for γ values, where the density in the spherical region behind the shock turns to zero at some value of ξ (Landau and Lifshitz, 1980).

The approximate solution to the strong explosion problem (Chernyi, 1957; Zel'dovich and Raizer, 2002) is apparently closer for the visual interpretation of the micro-explosion experiments. In this approach, the entire mass encompassed by the shock wave is concentrated in a thin shell behind the shock front. The density inside the layer is constant and equal to the density behind the strong shock (see Fig. 6.6). The mass of the shell is $M = \frac{4}{3}\pi R^3 \rho_0$. The thickness of the shell obeys the mass conservation law: $\Delta r = \frac{R}{3} \cdot \frac{\rho_0}{\rho_1}$. It is assumed that the velocity of the shell equals the velocity behind the shock, u_1, while the pressure driving the shell is the fraction α of the pressure behind the shock front, $p_c = \alpha p_1$ ($\alpha = 1/2$ from the condition that the total energy remains constant). The Newton law for the shell motion reads

$$\frac{d}{dt} M u_1 = 4\pi R^2 \alpha p_1.$$

The solution to the problem gives the radius and the velocity of the shock as follows:

$$D = a \cdot R^{-3/2} \tag{6.15}$$

$$R = \left(\frac{5}{2}a\right)^{2/5} . t^{2/5},$$

where a is the explicit function of the initial parameters:

$$a = \left(\frac{E}{\rho_0}\right)^{1/2} \left\{\frac{3}{4\pi} \frac{(\gamma-1)\cdot(\gamma+1)^2}{(3\gamma-1)}\right\}^{1/2} \tag{6.16}$$

6.2.5.1 Interpretation of the micro-explosion experiments using the strong point-like explosion solution

Let us take for the characterization of the equation of state for a solid sapphire, $\gamma = 3$. The initial energy of explosion $E = 100$ nano J; $\rho_0 = 4$ g/cm^3. Then with the major parameter, $E/\rho_0 = 0.25$ cm^5/s^2, one obtains $a = 0.564$ [cm$^{5/2}$/s], the distance, passed by the shock as function of time, $R = 1.147 \times t^{2/5}$, and the shock speed, $D = 0.564 \times R^{-3/2}$ at that distance (time in seconds, distance in centimetres). Thus, the shock reaches the distance $R = 400$ nm, where the shock is allegedly stopped (being transformed to the sound wave), ~7 ps after the shock emergence. The shock speed at this distance according to the solution is still 16.7 km/s because the cold pressure of material, p_0, has been not taken into account. One can see that the strong point explosion qualitatively reproduces the experimentally observed laser-produced structure (Fig. 6.6).

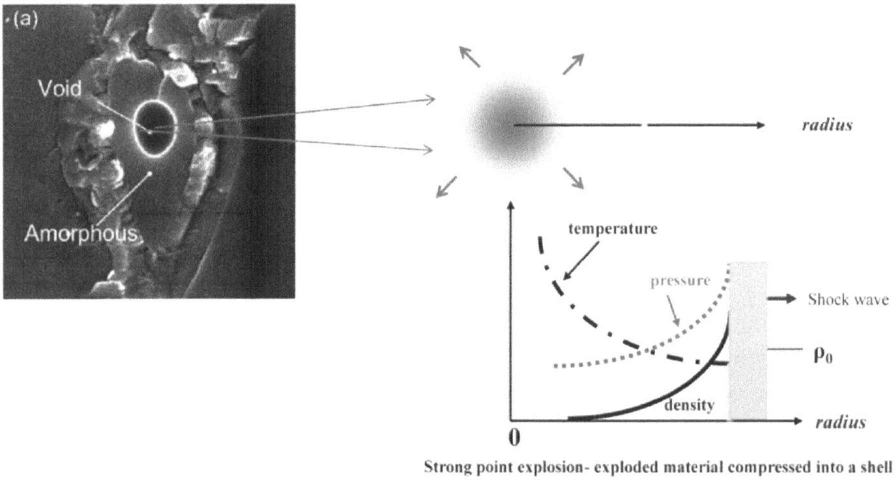

Figure 6.6 Structure produced by the micro-explosion (void and shell) qualitatively similar to that followed from the solution to the point-like strong explosion problem.

The shock gets converted into the acoustic wave, when the pressure behind the front becomes comparable to the cold pressure, more precisely, $p_1 \approx [(\gamma + 1)/(\gamma - 1] \cdot p_0$.

The characteristic length of the problem is, $r_0 = (E/p_0)^{1/3}$, the characteristic velocity is the sound velocity, $c_0 \approx (Y/\rho_0)^{1/2}$ and the characteristic time scale $t_0 = r_0/c_0$.

Let us estimate the characteristic scales of length and time relevant to the experiments in sapphire (Juodkazis et al., 2006), when 100 nJ were deposited inside the crystal. For sapphire, the cold pressure is approximately taken as the Young modulus, $Y \approx p_0$ = 0.4 TPa; ρ_0 = 4 g/cm^3. Then the characteristic distance is r_0 = 630 nm; $c_0 \approx 10^6$ cm/s; $t_0 \approx$ 63 ps. Note that the time count in these estimates starts after establishing the equilibrium when hydrodynamic motion begins. Let us point out that the structure produced by the confined micro-explosion only qualitatively resembles the point-like explosion in the perfect gas with the constant specific heats and density. In reality of the micro-explosion, the equation of state of a material is changing continuously during the process.

6.2.6 Summary of the Conditions after the Electron–Ion Energy Equilibration; Shock Wave Generation, Propagation and Stopping

The absorbed energy is confined in a volume with a radius of approximately 120 nm. In this area, the energy density is ~14 MJ/cm^3 (pressure 14 TPa), and aluminium and oxygen ions are separated at the distance of approximately 60 nm at the moment of the electron–ion temperature equilibration at the beginning of the hydrodynamic motion. The initial shock velocity is estimated to be around 60 km/s. The shock wave propagation time until conversion to the acoustic wave is estimated on the basis of hydrodynamics in a range of 10–60 ps. Thus, the whole material under effect of TPa pressure can be divided in two parts. In the central part, aluminium oxide was converted to the high-temperature solid density plasma where the light and heavy ions are spatially separated during the energy equilibration via the Coulomb collisions and following diffusion. The outer part, not affected by laser, experiences the passage of the strong shock wave driven by the pressure many times larger than the Young modulus of sapphire. Allegedly, whole material, which has been pushed out of the focal area, is embedded into surrounding pristine crystal in the form of the shell similar to the structure produced by the strong explosion in the perfect gas.

6.2.6.1 Analysis of the experimentally observed structure on the basis of the energy and mass conservation laws (Joudkazis et al., 2006; Gamaly et al., 2006)

The lateral and axial dimensions of the laser-modified region were measured with the scanning electron microscope (SEM). A focused ion beam milling was used to open the voids at their largest lateral cross section. The axial cross section of the laser-produced structure was obtained by cleaving the sample along the *c* axis. The observed pattern is shown in Fig. 6.7.

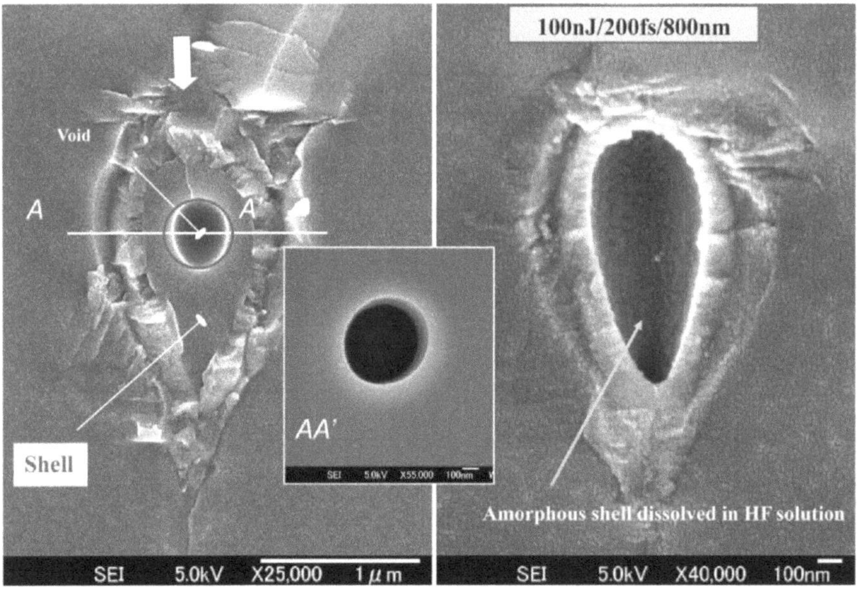

Figure 6.7 Scanning electron microscope images of the axial cross section of the laser-affected volume (left and right pictures) along with the lateral cross section in the centre (adapted from Gamaly et al., 2006).

It is clearly seen that the laser-affected region consists of a void surrounded by a shell twice thicker the void radius. This shell was identified as an amorphous sapphire. The amorphous sapphire can be completely dissolved in the hydrofluoric acid while a solid crystal remains intact. Indeed, the shell was dissolved in 10% aqueous solution of hydrofluoric acid and removed up to the smooth surface of the pristine sapphire in Fig. 6.7, right image. Therefore, the exact size of the laser-affected zone has been measured.

The outer surface of the shell apparently coincides with the surface where the shock wave was converted into the acoustic wave. The mass and energy conservation laws allow deducing the absorbed energy and shell density based entirely on the shell and void *post mortem* measurements. Indeed, at the surface where the shock wave transforms to the acoustic wave, the pressure behind the shock equals the cold pressure of a material. Therefore, the energy contained in the volume, embraced by this surface, is at least equal to the product of this volume by the cold pressure. Assuming that this volume is a sphere of the radius r_{stop}, and the cold pressure equals the Young modulus of the material, one estimates the low limit of the absorbed energy transformed into the energy of expanding and finally slowing down shock wave:

$$\frac{4\pi}{3} r_{stop}^3 \cdot Y \approx E_{abs} \tag{6.17}$$

The dependence of the diameter of the laser-affected zone has been measured as the function of the laser energy at the same focusing conditions (Juodkazis et al., 2006). Let us apply these considerations to the middle range of laser energy of 100 nJ, where the measured outer radius is $r_{stop} \approx 340$ nm. Then the formula (6.17) gives $E_{abs} = 54.9$ nJ thus giving the absorption coefficient of 0.55. It was found experimentally that at the energy of ~ 35 nJ, the void disappears; however, the laser-affected region extends to 250–300 nm. Energy considerations tell us that the absorption was lower in this case than at the high pulse energy and the pressure was lower than the cold pressure.

Let us apply now the mass conservation law for the estimation of the density of the shell surrounding the void. The whole mass contained initially in the void volume was driven out with the speed of a few tens km/s, compressed and embedded into a shell. The measured outer radius of the shell produced by 100 nJ laser pulse is $r_{stop} \approx 340$ nm, while the inner radius (void radius) equals $r_v \approx 160$ nm. Thus, the minimum estimate for the shell density increase only from the mass conservations (ignoring the compression by the shock wave) reads

$$\delta = \rho_{shell}/\rho_0 \approx \left[1 - \left(\frac{r_v}{r_{stop}}\right)^3\right]^{-1}. \tag{6.18}$$

The minimum density of the shell produced by the 100 nJ pulse thus is $\rho_{shell} \approx 1.116 \times \rho_0$. The experimentally observed dependence of the outer shell diameter and the void diameter on the laser energy has been fitted on the basis of the energy and mass conservation and using the measured threshold energy for the void formation (Fig. 6.8).

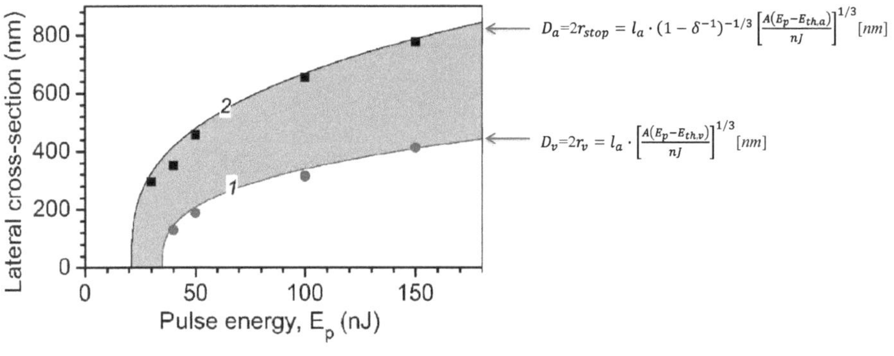

Figure 6.8 Diameters of the outer boundary of the shell (2, shock's stopping surface) and the void diameter (1) as the functions of the pulse energy, l_a = 100 nm; $E_{th,v}$ = 35 nJ; $E_{th,a}$=21 nJ; δ = 1.188; A = 0.6. Grey area is the thickness of the shell embracing the material ejected from the void and compressed by the shock (adapted from Juodkazis et al., 2006).

6.2.6.2 Evolution of the conditions and the structure of the laser-affected zone during expansion

It is instructive presenting the laser-affected area for the analysis as separated approximately in two parts. The first part is the energy deposition area (red spot in Fig. 6.9a) that increases remaining in the non-equilibrium state from the focal area to the end of the pulse up to twice-larger region during the temperature equilibration between electrons and ions and the heat transfer by the electrons (pink in Fig. 6.9a). Schematically, one can present the material state before the shock formation as a pressure jump while the material density remains constant (Fig. 6.9b). Conventionally such a pressure jump (discontinuity) breaks up into the shock and rarefaction waves. However, due to high electronic pressure and the spatial separation of the light and heavy ions, the shock front has a complicated internal structure and it is spatially spread. Simplifying the picture the shock compresses and heats the pristine crystal, while the rarefaction wave moves the plasma outside of the central area finally attaching it to the inner part of the shell surrounding the void (yellow in Fig. 6.9c). It appears that the spherical symmetry of explosion is maintained during the expansion as it follows from the final nearly spherical structure. Therefore, it is reasonable to suggest that the plasma part was converted to the inner layer of the shell, while the laser-unaffected crystal, compressed by the shock, constitutes the outer part of the shell bordering with the pristine crystal (blue in Fig. 6.9c).

The initial high pressure decreases during the expansion as $p \sim r^{-3}$ reaching the cold pressure of the crystal when the shock transforms to the sound wave.

The conditions for the material transformations in both parts of the shell are drastically different. The high-pressure shock affects the pristine crystal in a way similar to shock induced transformations (Trunin, 1994; Ahrens, 1980; Salleo et al., 2003). However, conditions in the plasma part are different. The pristine crystal is converted into high pressure/temperature plasma where the memory of the initial structure is lost. Allegedly, the light and heavy ion separation, produced during the energy equilibration process in plasma, holds during the stable central symmetric expansion. The discovery of the crystallites of bcc Al is the experimental proof of the ions spatial separation remaining during the cooling to the ambient conditions. One may conjecture that cooling from the disordered high entropy/pressure state allows some unusual paths (suppressed kinetic barriers) for the transformation to the unconventional final phase state.

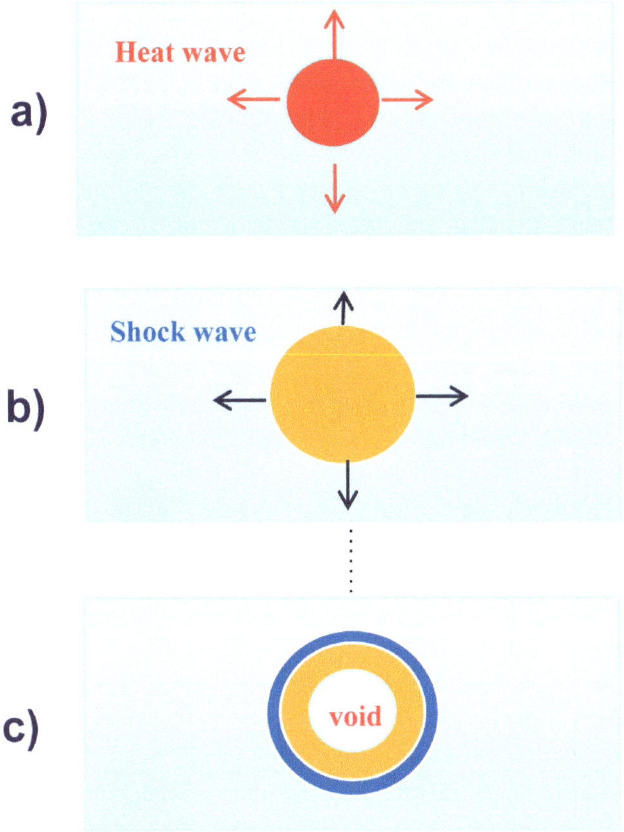

Figure 6.9 Schematic stages of the void and shell formation. At the end of the pulse all absorbed energy contained in electrons while ions are cold (red spot, a). During the energy transfer from electrons to ions and temperature equilibration the electrons transfer the energy on the twice-larger area than the focal volume (pink area, a). Shock wave emerges after the temperature equilibration (b). Shock compresses plasma into inner layer of the shell (yellow, c) and the pristine crystal in the outer part (dark blue, c).

6.2.7 Modelling the Hydrodynamics of the Confined Explosion (Juodkazis et al., 2006; Gamaly et al., 2006)

At the energy deposition stage, the processes of ionization, energy exchange and transport occur in the non-equilibrium three components plasma (electrons and two ions species) and should be described by the coupled set of kinetic equations and the Maxwell equations that is formidable task.

After the temperature equilibration, plasma attains the local thermodynamic equilibrium in a few picoseconds. The description of the mixture of ions in the frame of the "average ion approximation" reduces the problem to two temperature (electron–ion) hydrodynamics. The radiation hydrodynamics code "Chivas" has been used for the simulations of the confined explosion (Jacquemot and Decoster, 1991). The code is a one-dimensional, two-temperature (electrons and ions) hydrodynamic code accounting for the electron and ion thermal transport, electron–ion coupling, and transient ionization. The ionization states and opacity data were calculated assuming the local thermodynamic equilibrium. The equation of state implemented in the code is described in the reference (More et al., 1988). Three parameters, the mass density 2.2 g/cm^3, the bulk modulus Y = 75 GPa, and the binding energy of 3.29 eV/atom, define the equation of state for silica glass.

The explosion in silica glass has been calculated under assumption of the spherical symmetry. The absorbed energy in a range from 1 to 100 nJ was deposited homogeneously and instantaneously in the spherical volume of radius r_{dep} = 130 nm, which reproduces the energy deposition region after the temperature equilibration.

The simulations reproduce the initial stages of shock and rarefaction wave formation and beginning of the void formation. The calculated pressure at the deposited energy of 1 nJ in absorption region is comparable to the Young modulus and the void is absent. The instantaneous isochoric heating at the energy of 50 nJ produces the average ionization Z = 4.83 and the pressure of 2.6 TPa, well in excess the bulk modulus of silica (75 GPa). The strong shock wave emerges at the outer surface of the energy deposition sphere compressing a material to the maximum density twice of the initial one. The pressure behind the shock front rapidly decreases with the distance approaching the cold pressure at 50 ps. The shock wave transforms into the acoustic wave.

The formation of void is clearly seen. The asymptotic value for the compression ratio of $\delta = \rho/\rho_0 = 1.1$ was reached at 1 ns. It is in semi-quantitative agreement with the density of amorphous layer retrieved from the experiments, $\delta_{\text{exp}} = \rho/\rho_0 = 1.14$ (Juodkazis et al., 2006; Gamaly et al., 2006). However, the gas density in the central void region at 1 ns time is above 0.1 g/cm^3.

To conclude, the hydrodynamic calculations even with the simple model describe the shock propagation, stopping and the void formation during the period up to 1 ns in qualitative agreement with the experimental data.

6.2.8 Relaxation of the Laser-Affected Material to the Ambient Conditions

Let us define de-pressurization time as a moment when the pressure reduces to the cold pressure, i.e. the Young's modulus in a particular material. Assuming that the SW front velocity quickly reduces close to the sound velocity, one may estimate that time as $r_{\text{stop}}/v_{\text{sound}}$. Taking the speed of sound in sapphire $v_{\text{sound}} \approx 10^6$ cm/s, one obtains the respective time of ~ 40 ps. The average temperature at that moment can be estimated assuming that a material has a solid-state atomic number density ($n_a = 1.2 \times 10^{23}$ cm^{-3}) and the electronic recombination took place during shock-wave propagation consuming energy necessary for restoring the bonds of the order of the ionization potential ~10 eV. The heat capacity is also changing with temperature from 1.5 k_B as for an ideal gas to 3 k_B of a solid. Thus, it is reasonable to assume that the average temperature at the time when the shock wave stops is around 11,000 K, and the density is of the same order of magnitude as in the surrounding pristine crystal. The pressure at the moment of the termination of the shock wave is equal to the cold pressure of the surrounding pristine material, but the material is still in a partially ionized plasma state with the degree of ionization ~10%.

The isochoric cooling is a continuous process, but the rate of cooling has several stages, each attributed to a particular diffusivity. Each stage is characterized by a different diffusivity coefficient, which is temperature-dependent. The first stage continues during the shock wave propagation. In this stage, cooling from the hot dense plasma state is due to electronic heat conduction. The heat diffusion coefficient for the electronic heat transfer is $D = v_e^2/3v$. Here, $v_e \sim 2 \times 10^8$ cm/s is the electron velocity and the collision rate is $v_e \sim 5 \times 10^{15}$ s^{-1}, so $D = 2.7$ cm^2/s. Taking the shock-wave propagation distance as $L \sim 800$ nm and suggesting that the heat wave moves the same distance, the resulting cooling time is $t_{cool} = L^2/D \sim 1.8$ ns and the temperature at the end of the first stage equals ~1.5 eV, qualitatively comparable with the above estimate of the cooling during the shock deceleration and stopping. At 1.5 eV the electronic heat conduction is still efficient. One obtains $D \sim 0.24$ cm^2/s for the second stage when the heat wave propagates further into the bulk. The temperature drops down from 1.5 to ~ 0.19 eV (~2000 K) when the heat wave propagates over another 800 nm during 27 ns. The last stage is cooling by conventional linear heat conduction. The drop of temperature from 2000 to 300 K requires the increase of the cooling volume by 6.7 times, which means the heat wave propagates further by ~1.88 μm. Taking the diffusivity in SiO_2 as $D_{SiO2} = 4.76 \times 10^{-3}$ cm^2/s the time required for propagating this distance and cooling down to room temperature is ~5.2 μs.

Summing up, one concludes that the space and time scales of the full micro-explosion transformation solid-plasma-solid process constitute correspondingly ~2 μm and 5 μs. Beuton et al. (2017) studied theoretically and computationally the thermal, elastic and plastic modifications, following the cavity formation in fused silica. However, the time-dependent phase transformation at high-pressure stage still remains an unexplored area.

6.3 Interaction of the Gauss Beam Tightly Focused on the Surface of an Opaque Solid Buried under the Transparent Cover

The laser–matter interaction confined in the bulk of the transparent medium has two distinctive features in comparison to the laser-surface interaction. First, the whole absorbed energy is converted to the internal energy of the medium, expansion losses excluded. Moreover, the entire laser-affected material remains confined in the bulk of a medium making the pressure/temperature-transformed material available for the further studies. First experiments with the confined interaction were performed in the transparent materials (sapphire, silica, polystyrene). It has been understood that the confined interaction can be extended to the non-transparent materials if the opaque material is buried under the layer of transparent material and the laser is focused at the opaque surface through the transparent layer (Rapp et al., 2015). However, in this case, the formation of the extreme pressure/temperature conditions and a void with controlled shape becomes complicated. Different optical and material properties of both solids combined with the laser parameters impose many restrictions in comparison to the interaction with the homogeneous transparent crystal. Below, we analyse the experimentally studied case where the layer of silicon was covered by the layer of silicon oxide (silica) and the powerful laser beam has been focused at the silica–silicon interface (Rapp et al., 2015).

6.3.1 Upper Limit for the Energy Density Directly Delivered to the Opaque Medium through the Transparent Layer

In order to convert a dielectric to simple plasma (mixture of electrons and equal number of singly ionized ions), it is necessary delivering to electrons the energy in excess of the sum of the band gap and the binding (cohesion) energy, $\Delta_g + \varepsilon_b$. The binding energy responsible for the bonds in a solid is of the order of a few eV per atom (3–4 eV) for majority of dielectrics. However, the band gap for the opaque dielectric such as silicon is $\Delta_g \approx 1.1$ eV, while for silica it is around ~ 9 eV. The ionization threshold fluence for silicon, therefore, is sufficiently lower than that for silica. For that reason, the maximum laser fluence deposited directly into opaque material must be less than the ionization threshold in the cover the keeping the covering layer transparent. At this energy level, the influence of the processes in the transparent cover on the transformations in the opaque medium is excluded. The energy in excess of the ionization threshold deposited into the transparent dielectric converts it to plasma thus affecting the processes in the opaque medium through the heat conduction and the shock wave as we discuss later on. Let us first estimate the energy density limit for the experimentally studied case of silicon covered by the layer of silicon oxide (Rapp et al., 2015). Laser (170 fs, wavelength 790 nm, 0.1–1.0 μJ energy per pulse) was tightly focused (NA = 1.45) onto a Si-surface buried under a 10 μm-thick SiO_2 layer.

The scheme of laser focusing on the Si surface through silica layer is presented in Fig. 6.10.

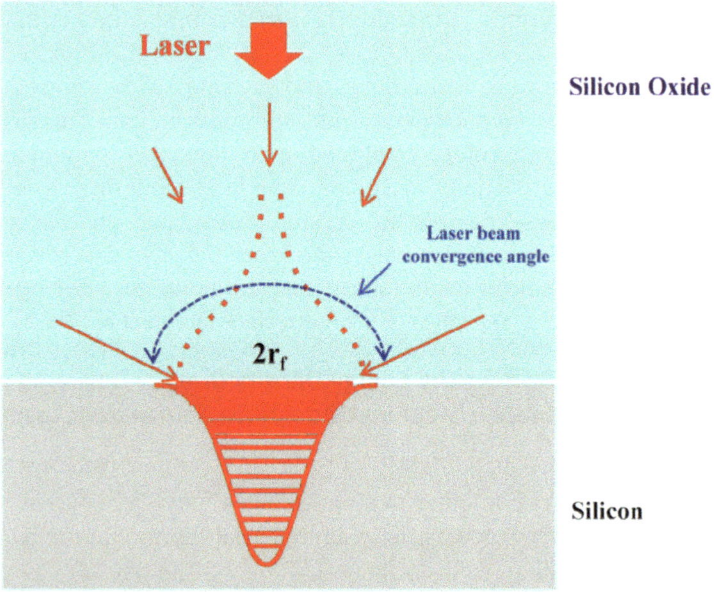

Figure 6.10 Laser interaction with silicon covered by the transparent silica layer at laser fluence higher than the ionization threshold in Si but lower than that in silica ($r_f \approx 3$ μm).

The red area in silicon in Fig. 6.10 depicts the absorption zone. The experiment with the energy density of $F \approx 2.6$ J/cm^2 (radius of the focal spot $r_f = 3.127$ μm), which is close to the ionization threshold in silica, was reported in Rapp et al. (2015).

Taking the absorption length in silicon of 30 nm, one gets the estimate for the energy density directly deposited in silicon that corresponds to the pressure of 660 GPa. This figure is four times higher than the cold pressure in Si, $\rho_0 \times v^2_{sound}$ =168 GPa as well as the cold pressure in silica of 54.9 GPa. Consequently, the strong shock wave propagates into silica. The relatively weak shock propagates into silicon. Both waves create an asymmetric laser-affected area at the interface. The silicon/silica interface of the laser-affected sample in this experiment has been sliced across in direction perpendicular to the boundary between two media by the ion beam milling. The photograph made with the scanning electronic microscope (SEM) reveals the void created by the laser (see Fig. 6.11).

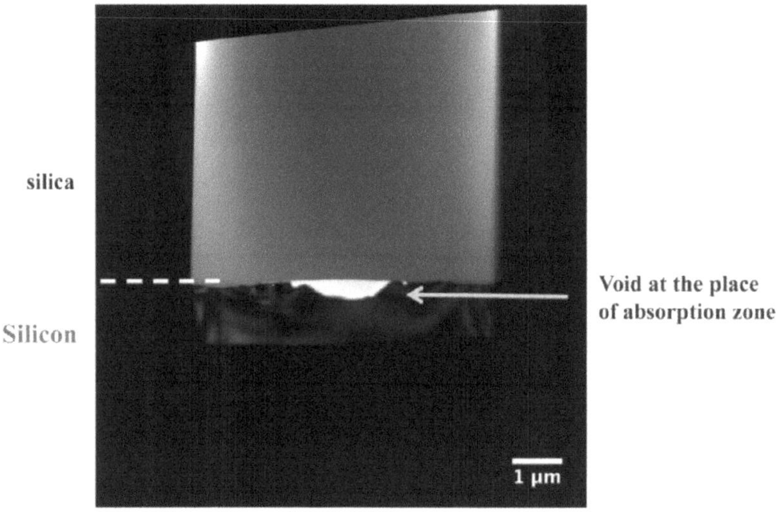

Figure 6.11 Cross section of the laser-affected area at the silica–silicon interface at fluence of 2.5 J/cm^2 (adapted from Rapp et al., 2015).

6.3.2 High-Energy-Density Interaction at the Silicon/Silica Interface (Rapp et al., 2015)

Let us consider the interaction at the extremely high energy density (800 nm, 150 fs, NA = 1.45; fluence 95–48 J/cm^2). After a couple of femtoseconds in the beginning of the pulse, silicon is ionized. In the next few femtoseconds, the ionization threshold in silica is achieved. From this moment, the laser light absorbs in the ionized silica.

The laser beam convergence angle in these experiments was bigger than in the experiments with sapphire. Therefore, it seems reasonable to estimate the energy deposition volume to the end of the energy transfer from electrons to ions making the assumptions similar to those made for absorption in sapphire. The energy deposition volume to the end of the pulse has a scale comparable with the limit for the energy concentration of the photonic beam set by the uncertainty principle, $\Delta x = \lambda/4\pi \times n$ (λ is the laser wavelength, n is the real part of the refractive index). For λ = 800 nm, and n = 36.38 nm, and the absorption length $l_{\text{abs}} \approx 30$ nm. The whole absorbed energy to the end of the pulse is confined in the electron sub-system. The major part of the absorbed energy is contained in silicon oxide, where two-three electrons are stripped from oxygen and silicon atoms (Gamaly et al., 2006). The electronic heat conduction in created plasma starts to be efficient during the pulse and after the end of the pulse. It leads to the smoothing out the electron temperature and the energy density over the absorption volume.

The hydrodynamic motion, leading to the void formation, begins after the energy transfer from the hot electrons to cold ions. The electron-to-ion energy transfer time $t_{e-i}^{\text{en}} \propto \frac{M_{\text{ion}}}{m_e} T_e^{3/2}$ grows up with the electron temperature increase. The electronic heat diffusion coefficient also increases at the high energy density, $D_e \propto T_e^{5/2}$. Electronic heat diffusion during the energy transfer time from electrons to ions thus significantly increases the area of the initial energy absorption to the moment when the hydrodynamic motion begins, $L_e \propto (D_e \cdot t_{e-i}^{\text{en}})^{1/2} \propto T_e^2$. Taking conservatively heat diffusion coefficient in plasma $D \approx 3$ cm^2/s, one gets the heat diffusion length during 3 ps equal to $L_e \approx (D \times t)^{1/2} \approx 30$ nm. It means that the heat diffusion homogenize the absorption volume making it close to the sphere of 30–40 nm diameter. A conservative estimate (ignoring any modification of the beam by the high-NA optics) gives the absorbed energy density in a range of 20 MJ/cm^3 suggesting maximum pressure in a range of 15 TPa and the electron temperature around a few hundred eV. In such conditions the electron–ion energy transfer by the electrostatic acceleration of ions by the gradient of the electronic pressure as well as by the Coulomb collisions are both responsible for the electrons—ions energy equilibration.

6.3.3 Shock Wave Generation in Silica Layer

The main scenario of the energy transfer from electrons to ions, generation of shock and rarefaction waves and the void formation in silica is similar to that described in the previous section. The atomic bonds become broken when the electrons deliver to ions the energy in excess of the binding energy. This occurs early during the energy transfer at the fluence larger than the ionization threshold, as in the discussed experiments. After electron–ion temperature equilibration, silicon oxide is transformed into the state of the high temperature, high entropy (fully randomized) solid density plasma. The mass difference between ions of oxygen (M_{Ox} = 16 a.au.) and Si (M_{Si} = 28 a.au.) is the reason that oxygen ions get the energy from electrons earlier than Si ions and start moving out of the energy deposition region forming the outmost front of the shock wave. Si ions start receiving the energy and moving later. Oxygen ions spatially separated from Si ions form the complicated shock wave front's structure. The difference in the ionization thresholds and the cold pressures values creates the conditions for formation of the asymmetric shock waves propagating into silica and silicon with the different velocities and at the different initial conditions before the shock front. This is clearly seen at the photo of the cross section of the laser-affected zone made by the ion beam milling across the silicon/silica interface under SEM (see Fig. 6.12, courtesy of Rapp et al., 2015).

The boundaries of the shock wave affected areas in silica as well as in silicon, clearly seen in Figs. 6.11–6.12, can be associated with the distances where the shock stops. Indeed, the shock practically stops when the pressure behind the shock front becomes equal to the cold pressure in the undisturbed material. Under the spherical expansion in homogeneous medium, this approach works well. However, for the clearly non-spherical case as in Fig. 6.12, the absorbed energy is disproportionately divided between the shocks propagating in the opposite directions from the main energy deposition area in silica. Note that the shock wave arrived into silicon from silica encountered the area ionized and heated early in the laser pulse time. Therefore, the conditions for the high-pressure phase transformations in pure silicon are different from those in silica. Indeed, pure silicon is first affected directly by the laser and then compressed by the shock coming from silica. However, silicon in silica under high pressure/temperature first spatially separates from oxygen and afterwards relaxes to the ambient from the disordered state.

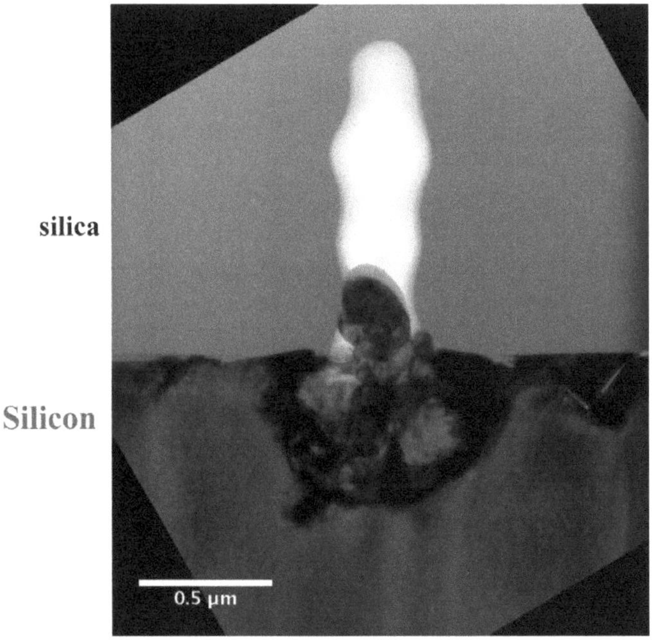

Figure 6.12 Cross section of the laser-affected area at the silica–silicon interface at fluence 95 J/cm^2 (adapted from Rapp et al., 2015).

6.4 Summary of the Gauss Beam–Generated Confined Micro-Explosions

The analysis of the experiments with the GBs gives the evidence that the energy density in a range of few TPa has been achieved. The limitations to the focal area as well as to the length of the focal volume set by the diffraction limit, making the focal volume, $V_{foc} \propto \lambda^3$. Therefore, an amount of laser-affected material in confined micro-explosion generated by the tightly focused

GB has a definite limit. The number of laser-affected atoms amounts a few picograms, making the structural studies of the laser-transformed material difficult.

6.5 Interaction of the Bessel Beam Focused inside a Bulk of the Transparent Crystal

Diffraction-free-like BBs create the focus many times longer than the Rayleigh length characterizing the extension of the GB focal volume. Therefore, application of the BB for the generation of the micro-explosions raised hopes for the transforming larger amounts of the material affected allegedly at as high pressure as it was achieved with the GB. Further, we describe the processes of the BB formation followed by the considerations of all the stages of the interaction process along with the analysis of the available experimental data.

6.5.1 Status of the BB–Transparent Crystal Interactions

It was demonstrated recently that the BB (150 fs, 2 μJ) focused inside sapphire crystal produced the cylindrical void of 30 μm length and 300 nm diameter (Rapp et al., 2016). The void volume, $V = 30\ \mu\text{m} \times \pi r^2 = 2.12 \times 10^{-12}\ \text{cm}^{-3}$, appears to be two orders of magnitude larger than that generated by the GB. The conclusions based solely on the void size measurements and on the energy and mass conservation laws without any ad hoc assumptions about the interaction process are the following (Gamaly et al., 2017; Gamaly et al., 2018). The material initially filled the void was expelled and compressed into a shell by the high-pressure shock wave. The work necessary to remove material with the Young modulus, Y, from volume, V, equals at least to $Y \times V = 0.848\ \mu\text{J}$ ($Y = 4 \times 10^5\ \text{J/cm}^3$—the Young modulus of sapphire). This is the evidence *of strong* (> 40%) *absorption* of the pulse energy. In order to generate a strong shock wave capable expelling such amount of material the absorbed energy should be concentrated in the central spike with the much smaller diameter than that of the void (still unknown either theoretically or experimentally).

6.5.2 Non-Diffractive Bessel Beam

It has been demonstrated by Durnin et al. (1987) that the monochromatic beam with an axial symmetry can propagate in a transparent medium with the space/time independent permittivity keeping unchanged transversal intensity distribution (being non-diffractive). The scalar diffraction theory works well when the wavelength is small compared with the aperture, and the propagation angle is not too steep (Born and Wolf, 2003). In this approximation, the spatial field distribution is described by the Helmholtz equation:

$$(\nabla^2 - k^2)E = 0;\ k^2 = \frac{\varepsilon\omega^2}{c^2} \tag{6.19}$$

In cylindrical coordinates, it reduces to the Bessel equation. The solution to this equation gives the electric field amplitude of the beam propagating in the z-direction conserving unchanged the transverse distribution:

$$E(r,z,k) = \exp(ik_z z) J_0(k_r r);\ k^2 = k_r^2 + k_z^2;\ k > k_r > 0 \tag{6.20}$$

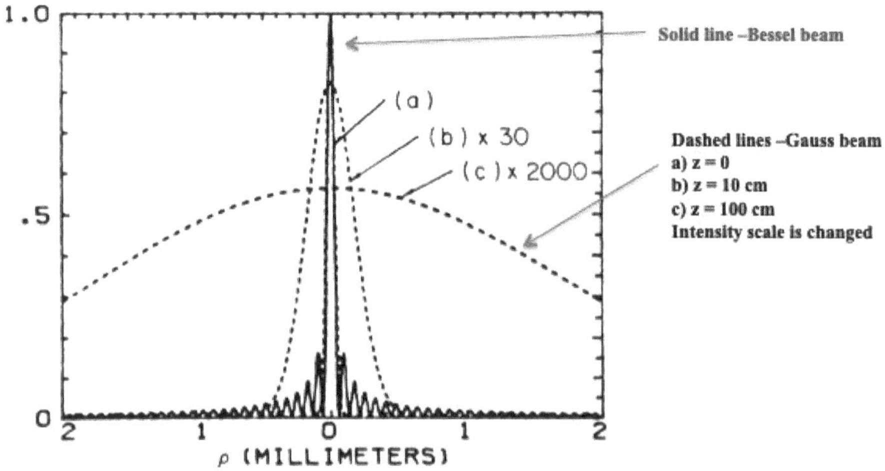

Figure 6.13 Comparison of the field distribution across the BB (solid line) and the diffracting GB (dashed lines) at the different distances from $z = 0$ (adapted from Durnin et al., 1987).

Here, J_0 is the Bessel function of the first kind zero order, which is limited at $r = 0$. These solutions are appropriate for free space (no boundaries or guiding surfaces) and for the dielectric permittivity of the medium constant in time and space without absorption and non-linearity. The intensity of the beam at $r > 0$ is proportional to $J_0^2 \propto r^{-1}$. Therefore, the total energy is infinite over entire space while the energy density is limited. The half-width of the central spike is $r_{\text{spike}} \propto k_r^{-1}$. The energy flux along the transverse direction in the diffraction-free BB is zero. The comparison of the electric field distribution across the beam for diffraction-free BB (at $z \geq 0$) and for the GB, which intensity decreases while propagating, is presented in Fig. 6.13.

6.5.3 Creation of the BB by the Circular Slit

The method for generation the intensity distribution in the focal zone having many features of the ideal BB with the finite aperture was proposed and implemented by Durnin et al. (1987). The circular slit of the diameter d and width Δd was placed in the focal spot of the lens with radius R. Each point along the slit acts as a point source, which lens transforms into a plane wave. The superposition of plane waves, whose wave vectors are evenly distributed over the surface of a cone, results in a diffraction-free transversal intensity distribution thus creating the monochromatic quasi non-diffractive BB. The tangent of the cone angle, θ, in accord with the geometrical optics, is defined as a ratio of the radius of the lens to the distance where the shadow zone begins (see Fig. 6.14). On the other hand, it follows from Eq. (6.20), $k_r = k \cdot \sin\theta$; $k_z = k \cdot \cos\theta$. The width of the slit shall comply with condition, $\Delta d << \lambda \times f/R$, in order to decrease the amplitude modulation of the diffraction envelope.

As it is seen in the inset in Fig. 6.14, the peak intensity oscillates with increasing amplitude and decreasing frequency until the point where the abrupt decline occurs. This point locates exactly at the beginning of the far right shadow zone. The parameters of the device producing the quasi-diffractionless BB in Fig. 6.14 allow defining the size of the focal zone and

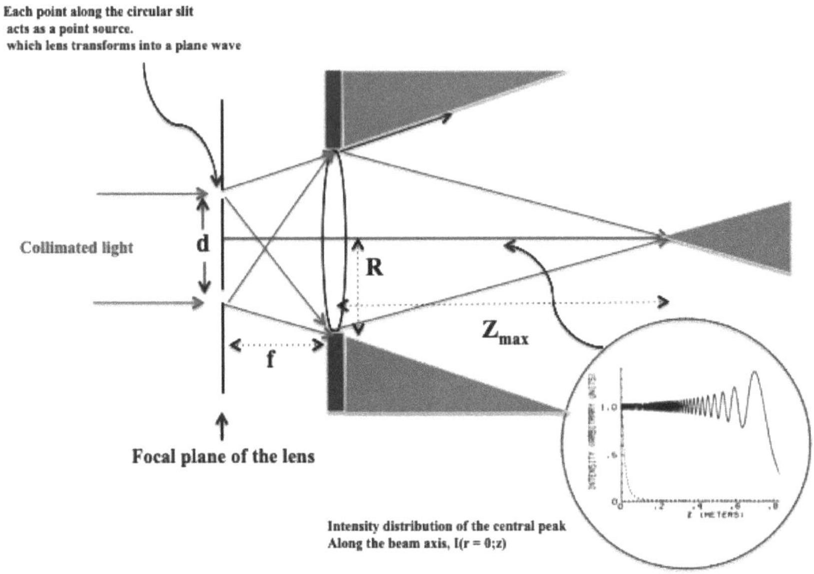

Figure 6.14 Scheme of the experimental set-up for the generation diffraction-free BB in air. Grey areas are the regions of the geometrical shadow. Insert at the right presents the calculated maximum intensity distribution (at $r = 0$) along the beam propagation (adapted from Durnin et al., 1987).

diameter of the central spike where the intensity has a maximum on the basis of the geometrical optics. Indeed, the distance from the lens until the beginning of the geometrical shadow is $Z_{max} = R/tg\theta$. The first zero of the zero order first kind BB locates at $r_0 \times k_r = 2.405$. The radial wave number $k_r = k \times \sin\theta$ (k is the laser wave number) allows to get the radius of the central, maximum intensity, spike, $r_{spike} = 2.405/k \times \sin\theta$. Note that *cw* laser has been used in these experiments at low intensity not affecting the optical properties in the medium of propagation (air). Our goal is using the BB for achieving the high energy density when a medium, where the beam focuses, is converted to plasma. Therefore, the intense BB interaction with the originally transparent medium is divided into two parts. The first period is the interaction with a medium before its transformation into plasma. Initial phase of interaction is focusing the incident beam inside a medium converting the intensity distribution in the focal volume to that close to the not-diffracting BB. Further, we consider the methods for creating the BB inside a medium at intensity not affecting the medium's optical properties.

6.5.4 Formation the BB by Transmitting the GB through the Axicon (or Spatial Light Modulator, SLM)

Let us consider the generation of the BB for the laser–matter interaction experiments aiming for creating the high energy density in the central core. The use of the conical lens (axicon) allows transforming the initial GB into diffraction-free-like BB of the limited aperture more efficiently that the circular slit.

The monochromatic (wavelength, λ) GB with the spatial intensity distribution across the beam, $A_0(r) = a_0 \times \exp\left[-\left(\frac{r}{r_0}\right)^2\right]$, is directed along the axis at the axicon with the wedge angle, α, and refractive index, n_a (see Fig. 6.15) The GB exactly behind the axicon transforms into the beam having the transverse distribution as follows: $B_0(r) = A_0(r) \times \exp(-ik_a r)$ (Jarutis et al., 2000). Here, the axicon's parameter is $k_a = 2\pi(n_a - 1)\alpha/\lambda$ (Fig. 6.15).

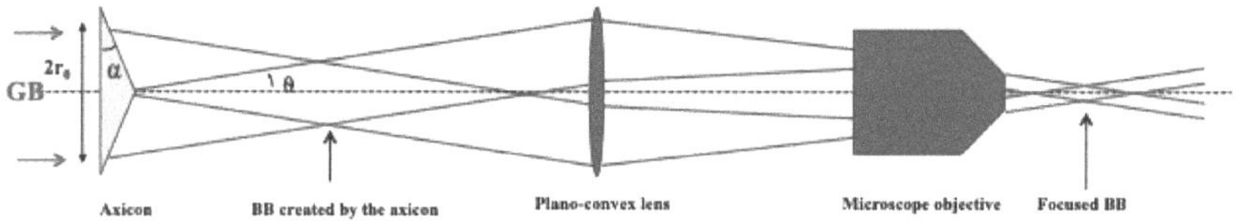

Figure 6.15 Transformation of the Gaussian beam (waist radius r_0) into the BB passing through the axicon (wedge angle α) and afterwards focused by the plano-convex lens and a microscope objective into a medium (courtesy Dr. Rapp).

The transverse intensity distribution of the beam transmitted through the axicon is $I(r) \propto J_0^2(k_a r)$. The length of the BB produced by the axicon depends on the axicon parameter, k_a, on the initial GB radius, r_0, and on the Rayleigh length in a medium, $Z_R = 2\pi r_0^2\, n_m/\lambda_0$, as follows (Jarutis et al., 2000):

$$L_a = \frac{Z_R}{k_a \times r_0} = \frac{n_m r_0}{(n_a - 1)\alpha} \tag{6.21}$$

For example, the axicon prepared from calcium fluoride (wedge angle, $\alpha = 1°$, $n_a = 1.4338$; $k_a = 2\pi \times (n_a - 1) \times \alpha/\lambda_0 = 0.0462 \times 10^4\ \text{cm}^{-1}$) transforms the GB ($\lambda_0$ = 1030 nm; r_0 = 1.925 mm) into the BB with the following parameters. The radius of the central high intensity region along the axis defined as the distance from the intensity minimum to the maximum at the axis is $r_{ca} \approx 2.405/k_a$ = 52.07 μm. The calculated length of the beam according to Eq. (6.21) is $L_a = r_0/\alpha \times (n_a - 1)$ = 254 mm, while measured length equals 210 mm (Rapp, L. 2020, private communication). Axicon also acts as the focusing device increasing the intensity of the incident GB. The ratio of the BB intensity to the intensity of GB is as follows (Jarutis et al., 2000):

$$\frac{I_{BB}}{I_{GB}} = \frac{2\pi^2(n_a - 1)\alpha}{\sqrt{e}} \frac{r_0}{\lambda} \max J_0^2(k_a r) \tag{6.22}$$

In the considered example, the axicon increases the intensity of the incident GB in 88.9 times.

Generating the BB with the spatial light modulator

The phase encoded on the SLM to generate the BB similar to that created by the axicon uses the same transfer function, $t(r) = \exp(ik\ (n_a - 1)a$. Therefore, parameters of the BB created by the SLM are the same as those generated by the axicon.

Focusing the BB inside the studied medium

For the further intensity increase the BB is focused into the studied medium. For example, the telescopic setup consisting of a plano-convex lens with a focal length of 200 mm and a microscope objective of ×60 (numerical aperture NA = 0.9) produces the BB with the length of $L_{Bf} \approx 103$ μm and the radius of the central core of $r_{cf} \approx 1.5$ μm. The transverse intensity distribution closely resembles the radial intensity distribution of the first kind zero order Bessel function with the maximum at the axis surrounded by concentric rings of gradually decreasing intensity (Dr. L. Rapp, private communication, 2020).

Comparing the volume of the central core of the focused BB to the focal volume of the tightly focused GB at low intensity not affecting the optical properties of a medium

Let us define the volume where the energy density is a maximum as the transverse area of the central spike by the focal length, L_{Bf}, of diffraction-free zone:

$$V_{\text{foc}}^{\text{BB}} = pr_{\text{cf}}^2 \times L_{\text{Bf}} = 728\ \mu\text{m}^3$$

It is instructive to present for the comparison the focal volume of the GB tightly focused in air as the area across of the beam's waist, d_w, by the doubled Rayleigh length, $L_R = 2\pi d_w^2/\lambda_m$, as follows (Juodkazis et al., 2006):

$$V_{\text{foc}}^{\text{GB}} \approx \pi^2 d_w^4 / \lambda_0 = \pi^2 \left(\frac{0.61}{NA}\right)^4 \times \lambda_0^3 \approx 0.356 \times \lambda_0^3$$

One gets $V_{\text{foc}}^{\text{GB}} = 0.389\ \mu\text{m}^3$ for $\lambda = 1030$ nm. Therefore, laser-affected volume in the interaction of the BB focused inside a crystal is more than three orders of magnitude higher than in the GB/crystal interaction.

6.5.5 BB Interaction with the Transparent Medium at Low Intensity

The difference between the non-diffractive BB being the solution to the scalar wave equation in the laser–matter interaction studies and the vector field equation approach can be clearly seen. The Bessel function of the first kind zero order can be presented as a sum of two Hankel functions of the first and second kind (Morse and Feshbach, 1953). It was indicated (Porras et al., 2004) that the formation of the non-diffracting BB could be considered as the interference of two conical running Hankel waves of first and second kind. One of the Hankel waves creates energy flow inward the radial direction, while the second wave creates the outward flow. In the non-diffracting BB both flows compensate each other making zero the resulting flux.

However, consideration of the axial-symmetric monochromatic electromagnetic wave E (E_r, E_z) in the absorbing medium produces the different results. Absorbed energy density per unit time/volume is the divergence of the Poynting vector:

$$Q = -\nabla \cdot S = \frac{\varepsilon_{\text{im}} \cdot \omega |E|^2}{8\pi}; \quad |E|^2 = E_r^2 + E_z^2 \tag{6.23}$$

The incident field structure near the axis is E (E_r, $E_\varphi = 0$; E_z); H (0; H_φ, 0). Hence, there are two components of the Poynting vector (energy flow), $\bar{S} = \frac{c}{4\pi}(\bar{H} \times \bar{E})$, along the z-axis and the radial direction:

$$S_r = -\frac{c}{4\pi}(H_\varphi \cdot E_z); \; S_z = \frac{c}{4\pi}(H_\varphi \cdot E_r) \; [\text{Joule/cm}^2\text{s}] \tag{6.24}$$

In the intense BB interaction with the initially transparent medium the ionization threshold is reached at the axis of the focal volume where the intensity is a maximum. It occurs early in the pulse time close to the beginning of the elongated focal region. Let us find the energy density when this conversion occurs.

6.5.6 Intense BB Interaction: Conversion a Dielectric into Plasma Near the Axis

From the condition for the ionization threshold, $\varepsilon_{\text{re}} = 0$, follows that the real part of the refractive index and imaginary part are the same, $n = k$, and the number density of conduction electrons is, $n_e \approx \varepsilon_0 \times n_c$, while the imaginary part of the permittivity expresses as, $\varepsilon_{\text{im}} = 2n \cdot \kappa$. Equation (6.23) reduces to the familiar form, $Q = \frac{2A \cdot I}{l_s}$, using connection the incident and evanescent field intensity through the boundary conditions at the surface, where $\varepsilon_{\text{re}} = 0$. The dimensionless absorption coefficient is estimated as in plane interaction. Later on, we estimate it in cylindrical geometry. The squared field expresses through the incident intensity, $|E|^2 = 8\pi I/C$ (linear polarization). We take the field penetration depth, $l_s = c/\kappa \cdot \omega$, as the transverse scale of the energy deposition region, $l_s/2$. Let us estimate the ionization threshold induced by 800 nm laser light ($\omega = 2.356 \times 10^{15}$ s^{-1};

n_c = 1.744 × 10^{21} cm^{-3}) in sapphire (ε_0 = 2.4; cross section for atomic collisions, $\sigma_{at} \approx 10^{-15}$ cm^2, velocity of electrons transferred through 10 eV band gap, $v_e \approx 2 \times 10^8$ cm/s). Then the electrons effective collision frequency follows, $\nu_{eff} \approx n_e \sigma v_e \approx \varepsilon_0 \cdot n_e \cdot \sigma_{at} \cdot v_e \approx 8.37 \times 10^{14}$ s^{-1}; $\varepsilon_{im} \approx 0.756$; $n = k = 0.615$; $A \approx 0.8$, and l_s = 207 nm. The energy density at the ionization threshold (using time averaged absorption coefficient and field penetration depth) is estimated as follows:

$$E_{abs} \approx \int_0^t Q d\tau \approx \frac{2A}{l_s} \cdot \int_0^t I^2 d\tau \equiv \frac{2A}{l_s} \cdot F(t) \tag{6.25}$$

In view of the importance of the absorption length, it is instructive to estimate this space scale from the different viewpoint.

The energy carriers, photons, focus to the axis. Therefore, the size of the area where the intensity has a maximum is set by the diffraction limit. However, there are no optical methods to define the limit in the conditions when the dielectric is converted to plasma. The uncertainty principle of quantum mechanics sets the ultimate limit for the spatial concentration of photons. Heisenberg's relation connects the photon's momentum and position:

$$\frac{\hbar\omega}{c} \times \Delta d \approx \hbar/2 \tag{6.26}$$

Therefore, the minimum size of the area where the photons with the frequency, ω, and wavelength, $\lambda = 2\pi c/\varepsilon_0^{1/2}\omega$ can be concentrated is $\Delta d \geq \lambda/4\pi$. For 800 nm light in sapphire (ε_0 = 2.4), Δd = 41.09 nm. The transfer of electrons from the valence band to the conduction band occurs under the action of intense high-frequency field. The amplitude of the electron's oscillations in such a field, $a_{osc} = eE/m_e\omega^2$, and the mean free path of the conduction electrons, $l_{mfp} = v_e/\nu_{ei}$, are both around a nanometre, that is much smaller than the diffraction limit. Therefore, the minimum size of the energy deposition region near the axis is of the order of the diffraction limit. We determine the ionization threshold from the condition that the absorbed energy is confined in the thermal energy of electrons transferred to the conduction band with the average energy equal to the band gap, $k_B T_e \approx \Delta_g$, $E_{th} \approx 1.5 \cdot \varepsilon_0 \cdot n_c \cdot k_B T_e$. Making equal the two expressions at the time, when the ionization threshold is achieved, one gets the threshold fluence from the equality, $E_{th} = E_{abs}$, as follows:

$$F(t_{th}) = \frac{3l_s}{4A} \cdot \varepsilon_0 \cdot n_c \cdot \Delta_g \tag{6.27}$$

At the laser/sapphire parameters as above, the absorbed energy density is 12 kJ/cm^3 while the threshold fluence for ionization varies in a range of (0.24–0.05) J/cm^2 depending on the transverse size of the energy absorption zone (207 or 41 nm). The laser fluence delivered to the end of the laser pulse for achieving high energy density should be in a range 10 J/cm^2 or more. Therefore, the transformation into the plasma nanowire along the axis occurs in the very beginning of the pulse starting the formation of elongated absorption cylinder. The elongation of the plasma cylinder continues in accord with the pulse propagation to the more distant part of a crystal (see Fig. 6.16).

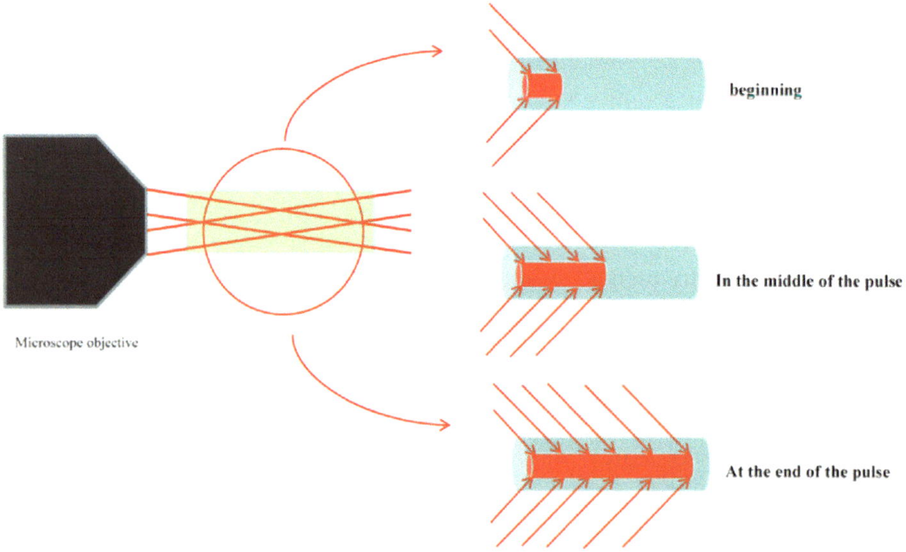

Figure 6.16 Formation of the cylindrical plasma core during the action of the short BB pulse. Plasma core is in red, pristine crystal is in blue. The boundary between two media is the surface where $\varepsilon_{re} = 0$.

Let us now consider the relation between the pulse duration, absorption, focal region length and laser-affected area diameter. Intense pulse converts initially transparent material into strongly absorbing plasma practically at the moment of its arrival at some space point. Therefore, the plasma region gradually increases along the axis as the pulse proceeds in time until the end of the pulse. The last portion of light arrives after travelling through the transparent crystal the distance $\sim t_p \times c/n$. The laser-affected distance then reads, $L_{las} = (t_p \times c/n) \times \cos\theta$ (θ is the cone angle). One can see now the difference between the BB-affected area in transparent medium (diffraction-free focus) and laser-affected area in the intense short pulse laser/crystal interaction. For sufficiently short pulses laser-affected area might be shorter than diffraction-free zone, $L_{las} < Z_{max}$. Thus, the laser pulse duration might be another lever (along with the cone angle) to control the energy deposition volume.

6.5.7 BB Interaction with Plasma

In order to describe the electric field inside the cylindrical plasma core 2D Maxwell equations with the space/time dependent permittivity should be solved, which is a formidable problem. Let us consider the beam with the electric field vector being in the plane of the incidence hitting the surface of the cylinder at the angle φ. The incident field structure near the axis is, E (E_r, $E_\varphi = 0$; E_z); H (0; H_φ, 0). The intensity along the axis (z-direction) in the non-diffracting beam does not depend on this coordinate.

A qualitative picture of the field behaviour in plasma can be obtained considering the cylindrical plasma as a conducting column in the same way as it has been done for the skin effect in the metallic wire (Tamm, 1976). Let us consider the space $r_0 \geq r \geq 0$ filled by plasma with the space-independent complex permittivity. The field directed along z-axis, E_z, drives the current. Therefore, we seek the solution to the Helmholtz equation in the cylindrical coordinates depending only on radius as in the solution (Durnin et al., 1997). The major difference is that the permittivity is a complex number with the negative real part, $\varepsilon = \varepsilon_{re} + i\varepsilon_{im}$. The axial field depends on radius and on the fast time, $E_z = E_{0z}(r) \times \exp(i\omega t)$. The Bessel equation has a familiar form:

$$\frac{\partial^2 E_{0z}}{\partial r^2} + \frac{1}{r}\frac{\partial E_{0z}}{\partial r} + k_r^2 E_{0z} = 0;\; k_r^2 = \frac{\varepsilon\omega^2}{c^2} = (\varepsilon_{\text{re}} + i\varepsilon_{\text{im}})k_0^2 \tag{6.28}$$

The solution to Eq. (6.28) limited at $r = 0$ is the Bessel function of the first kind zero order of the complex argument, $J_0(k_r r)$ (Morse and Feshbach, 1953). We consider the expansion of the solution into series near the axis where $k_r r \le 1$:

$$J_0(k_r r) \approx 1 - \left(\frac{k_r r}{2}\right)^2 + \frac{1}{4}\left(\frac{k_r r}{2}\right)^4 - \cdots \tag{6.29}$$

Taking the real part of the above expression, one obtains the radial dependence of the axial field as follows:

$$E_{0z}(r) \approx E_{0z}(r_0)\left[1 - \frac{\varepsilon_{\text{re}} k_0^2 r^2}{4} + \frac{(\varepsilon_{\text{im}}^2 + \varepsilon_{\text{re}}^2)k_0^4 r^4}{64} - \cdots\right]$$

In the considered case, $k_0 r_0 \le 1$ and $\varepsilon_{\text{re}} < 0$. Hence, the field at the axis at $r = 0$ is slightly lower than at the surface r_0, where a dielectric is converted to plasma. Let us estimate the axial velocity of electrons driven by this field with the current density, $j_z = e n_e v_e = \sigma E_z$. The conductivity conventionally expresses through the imaginary part of the Drude -like permittivity of plasma (see Chapter 3) as follows:

$$\sigma = \frac{\omega \times \varepsilon_{\text{im}}}{4\pi} = \frac{n_e v_{\text{eff}}}{4\pi \times n_{\text{cr}}(1 + v_{\text{eff}}^2/\omega^2)} \tag{6.30}$$

The electron's velocity reads

$$v_e \approx \frac{v_{\text{eff}}}{e \times 4\pi \times n_{\text{cr}}(1 + v_{\text{eff}}^2/\omega^2)}\left(\frac{8\pi I}{c}\right)^{1/2}. \tag{6.31}$$

Here, the electric field at the surface of the cylinder with the radius r_0 presented through the incident intensity on this surface, $E_z = \left(\frac{8\pi I}{c}\right)^{1/2}$. The light beam at 800 nm with intensity

5×10^{14} W/cm^2 drives electrons with the velocity 3.4×10^8 cm/s corresponding to the energy of 30 eV. In this estimate, we did not account for the contribution of the radial component of the field.

The modification of the electric field distribution along the radius after plasma formation can be qualitatively presented on the basis of the analytic solutions near the dielectric/plasma boundary (where the real part of the permittivity is zero) and in plasma by the modified Bessel function (see Fig. 6.17).

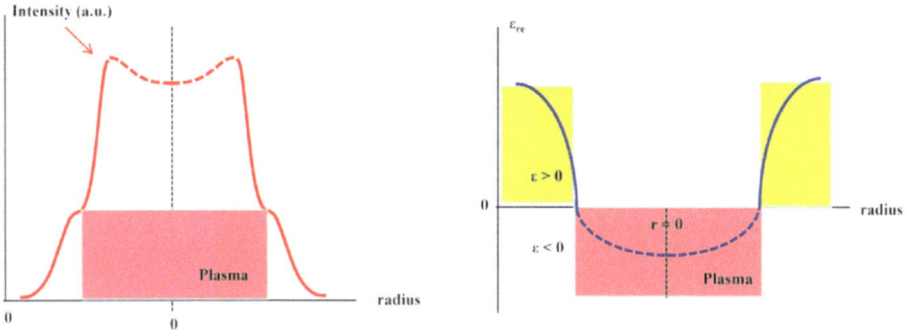

Figure 6.17 Qualitative picture of the intensity and permittivity radial distribution across the central core in plasma near the axis.

6.5.8 Energy Deposition in Plasma to the End of the Pulse

The intense BB with the pulse duration shorter of the major relaxation times deposits the energy into electrons of plasma filling the cylinder with the length approximately equal to the diffraction-free distance of Z_{max} and with the diameter of 40–60 nm, which is close to the diffraction limit/absorption length. The electron heat conduction transports the energy from the absorption area after the end of the pulse until the end of the energy transfer from electrons to ions. The electron-to-ion energy transfer time and the heat diffusion coefficient are both strong functions of the electron temperature.

The electron temperature at the ionization threshold is close to 10 eV at the absorbed fluence around 0.2 J/cm^2. In order to achieve high-energy concentration, the energy per pulse (and fluence) should increase 50 times. Correspondingly the plasma temperature, electron–ion energy transfer time and heat conduction time all increase. Electronic heat conduction flattens the temperature along the plasma cylinder levelling the energy density in the absorption volume before the beginning of the hydrodynamic motion. The electron-to-ion momentum transfer rate is a maximum at the ionization threshold, $\nu_{ei} \approx \nu_{max}$ (10 eV/$k_B T_e)^{3/2}$. We express the electron-to-ion energy transfer time and heat diffusion coefficient in sapphire through their values at the threshold and temperature as follows:

$$t_{ei}^{en} = (7.17 \text{ ps}) \cdot (k_B T_e/10 \text{ eV})^{3/2}$$

$$D_{diff} = (2.67 \text{ cm}^2/\text{s}) \cdot (k_B T_e/10 \text{ eV})^{5/2}$$

Then the distance travelled by the heat wave during the electron-to-ion energy transfer reads

$$L_{heat} = (D_{diff} \cdot t_{ei}^{en})^{1/2} = (4.37 \cdot 10^{-6} \text{cm}) \cdot (k_B T_e / 10\text{eV})^2. \qquad (6.32)$$

It is easily seen that the 50-fold temperature increase results in the smoothing of the temperature along a few tens of microns. At the same time, the heat conduction in the adjacent dielectric remains low thereby keeping the energy concentrated in thin cylinder along the axis. These estimates are very approximate giving rather qualitative than the quantitative picture. The more rigorous calculations of the field spatial distribution as well as the effects of non-ideal plasma should be accounted for. However, the experimental results are encouraging. A rather smooth, a few tens of microns long channel in sapphire is the experimental evidence supporting suggestion of nearly uniformly absorbed energy density distribution in the narrow plasma nanowire (Rapp et al., 2016).

6.5.9 Strong Cylindrical Explosion

Hence, it seems viable to consider the explosion induced by the BB deposited energy as a strong cylindrical explosion (Sedov, 1959). In this approximation, the size of the area where the energy density built up considered as being small in comparison to the distance travelled by the strong shock wave. The linear energy source is located along the line coinciding with the axis of cylinder. One-dimensional explosion takes place in the radial direction. Four initial parameters define explosion under adiabatic conditions (the entropy is conserved). These parameters are as follows: p_0, "cold" pressure in the medium; energy per unit length, E_{lin} (J/cm), in the case of cylindrical (line) explosion, density of a solid, ρ_0; and γ, the ratio of the heat capacity at the constant pressure to that at the constant volume. In general, two dimensionless variables, ξ, τ, along with two variables radius, r, and time, t, define the one-dimensional adiabatic explosion (Sedov, 1959):

$$\xi_{\text{cyl}} = r\left(\frac{\rho_0}{E_{\text{lin}} \cdot t^2}\right)^{1/4}; \ \tau_{\text{cyl}} = \frac{p_0 \cdot t}{(E_{\text{lin}} \cdot \rho_0)^{1/2}} \qquad (6.33)$$

The pressure in a medium finally decelerates and stops the shock wave thus determining the time scale of the problem through the second dimensionless relation from (6.33) as follows:

$$t_{\text{cyl}} \approx \tau_{\text{cyl}} \frac{(E_{\text{lin}} \cdot \rho_0)^{1/2}}{p_0} \qquad (6.34)$$

The cylindrical 30 μm-long channel was produced in sapphire by the laser pulse with the energy, E_{las} = 2 μJ (Rapp et al., 2016). The linear energy density delivered in this experiment constitutes, $E_{\text{lin}} = 0.67 \times 10^4$ J/cm. Initial mass density and "cold" pressure of sapphire are respectively, ρ_0 = 4 g/cm^3 and p_0 = 0.4 TPa). The time scale of shock propagation and stoping using these parameters can be estimated in accord with Eq. (6.34) as $t_{\text{cyl}} \approx 40$ ps.

The exact solution to the problem has been obtained in the self-similar case of the strong explosion when pressure of the medium was neglected (Sedov, 1959). Solution in this case depends only on two dimensionless quantities, γ and ξ. This solution allows determining the shock front position and velocity as the function of time. The position of the cylindrical shock front is expressed from (6.33) using $\xi_{0,\text{cyl}}(\gamma)$ obtained from the self-similar solution:

$$R = \xi_{0,\text{cyl}}(\gamma)\left(\frac{E_{\text{lin}}}{\rho_0}\right)^{1/4} \cdot t^{1/2} \tag{6.35}$$

The shock front velocity is

$$D = \frac{dR}{dt} = \frac{1}{2}\frac{R}{t} = \frac{1}{2}\xi_{0\text{cyl}}\left(\frac{E_{\text{lin}}}{\rho_0}\right)^{1/4} \cdot t^{-1/2} = \frac{\xi^2_{0,\text{cyl}}}{2R} \cdot \left(\frac{E_{\text{lin}}}{\rho_0}\right)^{1/2} \tag{6.36}$$

Accounting for the dependence $\xi(\gamma)$ in the spherical explosion results in an increase in the numerical coefficient by five times. Taking a similar correction in the cylindrical case of (6.36) gives $D \approx \frac{1}{R} \cdot \left(\frac{E_{\text{lin}}}{\rho_0}\right)^{1/2}$.

For cylindrical explosion in sapphire with the parameters as above, one gets the shock wave's speed $D \sim 40.9$ km/s at the distance from the axis, $R = 100$ nm. At the distance of 400 nm, the shock wave velocity decreases down to 10 km/s, which is close to the speed of sound in the cold sapphire. Near this distance, the conversion of the shock into the acoustic wave occurs.

Thus, considering the BB induced cylindrical channel as produced by the linear strong explosion gives reasonable semi-quantitative estimates for the shock wave velocity of a few tens km/s.

6.5.10 Experimental Evidence of the High Energy Density Achieved by the BB Induced Cylindrical Explosion

The BB (800 nm, 150 fs, 2 μJ), focused inside sapphire, produced the cylindrical void of 30 μm length and 300 nm diameter (Rapp et al., 2016). The void volume, $V = 30\ \mu\text{m} \times \pi r^2 = 2.12 \times 10^{-12}\ \text{cm}^{-3}$, appears to be two orders of magnitude larger than that generated by the GB with the energy per pulse 20 times less (Joudkazis et al., 2006).

These experiments confirmed that the BB creates the high-pressure in the larger mass of a material than in the GB generated micro-explosion. The analysis based solely on the void size measurements and on the energy and mass conservation laws allows making conclusions on the laser absorption and indirect indication on the high energy density achieved (Gamaly et al., 2017, 2018).

The material initially filled the void was expelled and compressed into a shell by the high-pressure shock wave allegedly in the cylindrical strong explosion. The work necessary to remove the material under the "cold" pressure, p_{cold}, from a volume V equals at least $p_{\text{cold}} \times V = 0.848$ μJ. We take the cold pressure equal to the Young modulus of sapphire, $p_{\text{cold}} = \rho_0 \times (c_{\text{sound}})^2 \approx Y = 4 \times 10^5$ J/cm^3. Thus, the ratio of the energy spent for the removal of material from the void to the total laser energy is $(p_{\text{cold}} \times V)/E_{\text{las}} = 0.424$. Therefore, at least 42% of the pulse energy was absorbed. The analysis in the previous sections demonstrated that in accord with the hydrodynamics a strong shock wave capable of expelling such amount of material should be generated in the central spike with the much smaller diameter than that of the void. Taking conservatively $d_{\text{spike}} \approx 100$ nm one estimates that at the absorbed energy density the maximum pressure of 3.6 TPa was achieved.

The experiments that gave evidence of the shock wave movement predominantly in the radial direction (the strong cylindrical explosion) were described in the reference (Wang et al., 2017). The multiple BBs produced in the plastic (PMMA) many cylindrical channels *with the open ends* by the single BB pulse action. The sample has been weighted before and after the laser action with the high accuracy. The authors found that the weight's difference of the sample before the experiment and weight the sample with the multiple channels produced was the same within a few percents accuracy. This result is the evidence that the practically whole material ejected from the cylindrical void was moved predominantly in the radial direction, compressed and placed on the wall of the cylinder.

The analysis of the experiments, producing $L = 30$ μm long, 300 nm wide, cylindrical channels in sapphire, supports the idea of the strong cylindrical explosion with the predominant radial material motion and compression. It is easy to see that if such explosion produces cylinder with the open ends the material ejection along the axis during the shock wave propagation and stopping is small. Indeed, the radial shock wave effectively stops at the radius of r = 400 nm. Assuming the same axial speed as the radial one, the material at the length $2r$ = 800 nm would be ejected from the both ends open. This exaggerated estimate suggests that only $(2r/L)$ = 0.027 fraction of the laser-affected material from the channel would be ejected from the open ends. This figure is close to the weight difference measured in the experiments (Wang et al., 2017).

6.6 Solid-Plasma-Solid Transformation in Confined Micro-Explosion: New Path for High-Pressure Material Phase Formations

Conventionally the transformations of different materials at high temperature and pressure are studied at stationary and dynamic conditions. The diamond anvil cell creates static pressures up to the limit set by the diamond Young modulus of 1 TPa. The pressure of 364 GPa and 5,500°C in the sample achieved in the lab recreated the conditions at the Earth's core. The pressures well over TPa and high temperatures have been created behind the strong shocks generated by the explosives and mega-joule-class lasers. In the static condition, the pressure in a material slowly increases in such a rate that material remains all the time in thermodynamic equilibrium while atoms are moved into the different spatial positions of new phase. In the dynamic methods, the strong shock wave delivered to the sample affects directly atoms (ions) of the initial structure. A material (atoms (ions) behind the shock front is compressed and heated by the shock. The high-pressure/-temperature/-density conditions in the shock-affected material last for a short period until the pressure release. Finally, after the pressure release, the pressure-affected material is dispersed. Thus, in both static and dynamic methods, the atoms in the initial spatial positions are subject to either slow or fast exposure to the extreme pressure/temperature. Electrons and ions are in equilibrium in both cases.

The conditions created by the confined explosion induced by the intense short laser pulse action are different. First of all, the massless carriers deliver the energy into the electronic sub-system within a volume, located inside the pristine material, while the ions remain in the vertices of the initial crystal. Electrons are swiftly heated to the temperature well in excess the binding energy and ionization potentials reaching the energy density of a few MJ/cm^3. Electrons transfer the energy to ions via the multiple Coulomb collisions converting a crystal into chaotic high entropy plasma state where the memory of the initial crystalline state was completely lost. Materials allowing focusing a laser beam in the bulk should be transparent at low intensity. Usually these are oxides (experimentally studied aluminium oxide, Al_2O_3, silicon oxide, SiO_2) consisting of heavy (Al, Si) and light (oxygen) atoms. The mass difference leads to the spatial separation of heavy and light ions during the electrons-to-ions energy transfer via the Coulomb collisions: light ions receiving the energy faster and start to move earlier in time than the heavy ones. Note that the spatial separation of the heavy and light ions further distances the ions positions from those in the initial crystal. The spatial separation strongly depends on the electrons' temperature (higher the temperature, longer the separation distance) and may achieve hundreds of nanometres. The ions pressure is a maximum at the moment of the electron–ion temperature equilibration, while the density is still the same as the initial one. The state of a material in a region, where the energy has been absorbed, is a starting point for returning to the ambient conditions. The whole pressure/temperature affected material remains confined inside a pristine crystal. The state of this material is follows: Solid density plasma at the temperature of around 100 eV, pressure of a few TPa with the heavy and light ions spatially separated. The memory of the initial state is lost. This state might be compared to the "primeval soup" in the plasma core of stars and the young planets, however short lived. Hydrodynamic motion begins when the ions got the energy. The strong shock wave propagates into the cold material surrounding the absorption zone affecting a pristine crystal by the high-pressure action in the same way as in the dynamic experiments described above. However, the whole shock-affected material remains confined as the part of the compressed shell inside the bulk of pristine crystal. The expansion of the plasma regions proceeds in a mode of point (or linear in the case of the BB interaction) strong explosion compressing the plasma of the absorption zone into a shell adjacent to the material compressed by the shock. Therefore, the confined explosion has two zones where the transformation of material under the action of high pressure occurs at the different initial conditions.

(1) The first zone is the absorption region transformed into the high entropy hot solid density plasma. In this zone the electronic temperature and pressure reaches a maximum to the end of the laser pulse. During the energy transfer from electrons to the ions of the different mass, the spatial separation of the light and heavy ions occurs. This zone becomes the pressure source after the energy transfer to ions leading to the generation of the expanding shock wave propagating into the pristine crystal. Further evolution of the absorption zone occurs during the rarefaction wave development resulting in the condensation of plasma into the solid shell surrounding the void.
(2) The second zone is the pristine crystal compressed by the strong shock expanding from the absorption zone into the shell. The outer surface of this shell is adjacent to the pristine crystal, while its inner surface borders with the absorption zone plasma converted by the explosion into the solid layer.

It is seen from the comparison with the experiments that the qualitative and semi-quantitative description of the plasma state before the beginning of the hydrodynamic motion can be reasonably presented. However, the structural transformations of the material under high-pressure action, following cooling and return to the ambient conditions, are still unclear because the conditions for the phase transformation are unusual.

6.6.1 Discovery of Superdense bcc-Al in Sapphire Irradiated by the Tightly Focused Gauss Beam

In the experiments of Vailionis et al. (2011), 150 fs laser pulses were focused in α-Al_2O_3 crystals (corundum, a variety of sapphire) at the depth of several micrometres. Sapphire has a band-gap of 9.8 eV, and the bulk modulus of B = 239–257 GPa (Smyth et al., 2000). The void and the shell inside the sample of the laser-affected sapphire are well discernible by an optical contrast and well defined. In contrast, it is difficult to discern the compressed shell boundaries in the laser-produced micro-explosion in silica or polystyrene.

The laser-affected zone was opened by the focused ion-beam milling. Then the transmission and scanning electron microscopes (TEM and SEM) were used for measuring the shell thickness, Δr, and the average radius of the void, r_{void}, both being $r_{void} \approx \Delta r \approx$ 180 nm.

Synchrotron X-ray diffraction microanalysis (μXRD) has been used for searching possible high-pressure phases. Surprisingly the μXRD provided the explicit evidence of the new bcc-Al phase predicted long ago but never observed experimentally. The observed formation of the crystalline aluminium out of aluminium oxide is the experimental evidence of an unusual and unexpected phenomenon, namely spatial separation of oxygen and aluminium in the high-temperature solid density plasma. However, the analysis provided above demonstrates that in the high-temperature non-equilibrium plasma, where the electrons are swiftly heated while ions remain cold, the Coulomb collisions transfer the energy to light ions much faster than to the heavy ones resulting in the ions spatial separation. The discovery of bcc-Al is the evidence that this new phase state was created in the originally high pressure/temperature region of the solid density plasma (absorption zone) during its cooling to the ambient conditions.

6.6.2 Discovery of New Tetragonal Polymorphs of Silicon

Single laser pulses (170 fs, 790 nm, 0.1–1.0 μJ) have been tightly focused through the 10 mm-thick amorphous silicon dioxide layer (SiO_2) at the interface with the adjacent opaque crystalline silicon layer (Rapp et al., 2015). Electron diffraction studies of the laser-affected region revealed several metastable end silicon phases. We described the complex micro-explosion generated near the interface between transparent and opaque material in Section 6.3. The laser beam hits silicon layer first producing ionization and small energy deposition in it. A few femtoseconds later, the ionization of silica (silicon oxide) occurs, and then the major part of the absorbed energy is deposited in silica. Therefore, the absorption zone where the pressure and electrons temperature are maximum is located in SiO_2. In this zone, the silicon oxide is converted into non-equilibrium disordered plasma with hot electrons and cold ions to the end of the pulse. The energy transfer from hot electrons to ions occurs after the end of the pulse via the multiple Coulomb collisions. The light oxygen ions got heated first and move apart from the heavy Si ions becoming spatially separated on the space scale of hundreds of nanometres before the silicon ions are heated. After electrons and ions temperature equilibration, the ions start moving generating the strong shock wave (SW) propagating into cold silica and to the slightly heated silicon. Therefore, new phases of silicon may be formed either in the energy deposition zone where silicon became spatially separated from oxygen or behind the strong SW propagating in pure silicon. All material ejected via the strong explosion from the void and compressed into the shell may contain the new phase states of high-pressure transformed silicon.

The predominant feature of the conventional static and dynamic methods is that an initially crystalline arrangement of atoms is subjected to either slowly increased pressure in the DAC or rapidly applied SW pressure. Under such conditions, the atoms are forced into closer inter-atomic positions and induce transformation from one crystalline state into another more dense state.

The theoretical studies demonstrated that diverse material structures could spontaneously organize themselves during isochoric cooling from hot dense plasma composed of a completely disordered mixture of elements. It is likely that fast cooling makes more energetically accessible transition to new structures allowing overcoming the kinetic barriers. Note that the micro-explosion created plasma cools down isochorically to the ambient temperature with an unprecedented quenching rate of up to 10^{14} K/s (Juodkazis et al., 2006).

Such an experimental path to new structures starting from a completely disordered state is in close analogy to the computational approach ab initio Random Structure Searching (AIRSS) (Picard and Needs, 2010, 2011). Transition to novel phases under high pressure, calculated in AIRSS, starts with the disordered material. In this approach, crystal structures are generated through the repeated full relaxation of randomly created initial structures, within the density functional theory (DFT). AIRSS calculations, for example, have led to the prediction of incommensurate aluminium with an unconventional spatial distribution of atoms and valence electrons (Picard and Needs, 2010).

Electron diffraction studies of (Rapp et al., 2015) revealed the presence of bt8 and st12 silicon structures, which have been predicted theoretically previously, but have not been observed in nature or in laboratory experiments. Therefore, the analysis in Section 6.3, the ab initio calculations and the experimental results suggest that it is highly likely that the new structures of silicon were formed within laser-absorption zone in silica where the spatial separation of Si and oxygen ions occurred in completely randomized plasma. In addition, the presence of the as yet unidentified silicon phase, Si-VIII and two of other predicted tetragonal phases is highly likely within laser-affected zones.

In this study, we concentrated on the methods for the generation, identification and control of the extreme pressure/temperature conditions created during the confined micro-explosions inside the bulk of a crystal. Micro-explosion produces a void inside a crystal when the absorbed energy density exceeds the cold pressure (strength) of a material. Therefore, applications of the micro-explosions are twofold. Here, we concentrate on the achievement of the extreme pressure for searching the novel material structures, which might be created in the unusual conditions of confinement explosion following by the isochoric cooling to the ambient conditions. However, the formation of voids accompanying the explosion is of interest for the micro-machining for the formation of complicated 3D structures within a transparent material. The presented analysis indicates the ways for controlling the size of the voids produced.

6.7 Summary: Confined Micro-Explosion by the Gauss and Bessel Beams

Confined micro-explosions by the GB and BB are two different areas of studies and applications.

(I) First, it has been demonstrated that tightly focused GB and BB produce a pressure in a range of a few TPa exceeding the strength of any existing material. New materials—bcc Al, novel phases of silicon—were produced in the novel solid-plasma-solid transformation mode inherent to the confined micro-explosion. The novel phases were conserved in the bulk of pristine crystal.

- Tight focusing of the GB close to the diffraction limit with the energy of 100 nJ per pulse allows achieving the pressure exceeding a few TPa:

$$P \propto \left(\frac{E_{\text{las}}}{\lambda^3} \right) > TPa$$

- The use of the tightly focused BB allows an increase by a couple of orders of magnitude the amount of high-pressure affected material keeping the maximum pressure in excess of a few TPa

(II) Second, the area of the studies and applications relates to the use of the tightly focused GB and BB for micro-machining producing different 3D structures inside transparent materials

- The radius of the void produced in the transparent medium by the GB is estimated by scaling through the energy of the beam and the density and the speed of sound of a material:

$$r_{v,\text{sph}} \propto \left(\frac{E_{\text{las}}}{\rho_o v_{\text{soud}}^2} \right)^{1/3}$$

- The radius of the cylindrical channel produced by the BB focused in the medium with the length of the focus

$Z_{\max}$ >> L_R, defined by the parameters of the optical system, is as follows:

$$r_{v,\mathrm{cyl}} \propto \left(\frac{E_{\mathrm{las}}}{Z_{\max} \cdot \rho_o \cdot v_{\mathrm{soud}}^2} \right)^{1/2}$$

Conclusions and Future Directions

The book is dedicated to the concise description of the material transformations in solids by the laser pulses shorter of the major relaxation times. The subtle modifications at low intensity up to the creation of the extreme energy density at the laboratory tabletop are considered in succession.

We demonstrate that at the intensity below the damage threshold, one can excite coherent phonons and create the unusual transient states of matter with the mixed metal/dielectric properties controlled by the laser and material properties.

The intense ultra-short laser tightly focussed inside a transparent dielectric generates the high energy density in a sub-micron-size volume and creates a void. The experimental and theoretical studies established two facts: (1) the absorbed energy density reaches over a few TPa level; (2) the shape and size of the void can be controlled. These two findings defined two directions of further studies and applications: materials transformations under the high pressure at the laboratory tabletop and creation of complicated three-dimensional structures within the transparent dielectrics—micro-machining transparent materials for application in photonics.

The unique features of the material conversion during the confined micro-explosion are the following:

- The record-fast heating creates the high-pressure/temperature solid density plasma (akin a short-lived "primeval soup" of stars).

Femtosecond Laser-Matter Interactions: Solid-Plasma-Solid Transformations at the Extreme Energy Density
Eugene Gamaly

ISBN 978-981-4877-40-4 (Hardcover), 978-1-003-25661-8 (eBook)
www.jennystanford.com

- The record-fast cooling converts plasma back to solid being confined inside a pristine crystal at the ambient conditions.
- The previously unobserved high-pressure phases of aluminium and silicon were identified by x-ray and electron diffraction at ambient conditions (Juodkazis et al., 2006; Rapp et al., 2015).
- The novel structures are the result of the phase transformation along the unconventional solid-plasma-solid path when the kinetic barriers were apparently lowered.

The laser intensity during the interaction increases up to hundreds of TW/cm^2 changing the material properties from solid to plasma on the time scales of femtoseconds and space scale of nanometres. Consequently, the first five chapters are dedicated to the description of the ultra-short laser–matter interactions and material transformations basics in the non-equilibrium conditions. We emphasize the differences between the material transformations in equilibrium and non-equilibrium based on the fundamentals of plasma, optics, and the solid-state physics with glimpses from the quantum mechanics. The methods and processes leading to the extreme energy density are then considered in Chapter 6 along with the experimental results evidencing the formation of previously unobserved novel material phases.

Summing up, there is the general understanding of the major processes leading to the extreme energy density. However, the studies revealed the new problems, which need understanding and solution. Those problems are listed below.

Change the internal structure of the intense beam focusing by the high-NA optics

The major process is the focussing of the incident beam to the diffraction limit in a transparent crystal. This process has been studied theoretically and experimentally only at low intensity when the optical properties of a medium remain unaffected by the laser field.

It has been found that the high-NA focussing of the Gauss beam changes the internal structure of the field in the focus, shape of the focal zone, and orbital moment of photons and probably induces the Berry phase. It is unknown how these changes influence the interaction process. Nevertheless, the experiments with the tightly focussed Gauss beams and with the intense Bessel beams evidence the absorption of at least 40% of the beam's energy.

Ionization and band gap collapse in dielectrics

The main process changing the optical properties of dielectric in the intense electric field is ionization. Semi-quantitative calculations produce a qualitative description of the processes in the absorption zone. Kinetics of ionization of the wide-band-gap dielectrics in space and time still does not have the proper theoretical description. The time evolution of the conduction electrons' distribution function, band gap shrinking, and collapse are still far from full understanding.

Effect of the evolution of the statistical distribution on the relaxation to the ambient conditions

The distinctive feature of the ultra-short laser–matter interaction is that the energy distributions of electrons and ions are far from equilibrium and are modifying during much longer period than the pulse duration. However, *the kinetic effects of the evolution of the distribution function on the phase transformation and relaxation* to the ambient conditions are unknown.

The light and heavy ion dynamics in the non-equilibrium plasma

The experiments revealed that the light and heavy ions in the non-equilibrium plasma become spatially separated. *However, the kinetics of the spatial separation and transition to hydrodynamics accompanied by the new phase formation is still in dark.*

Novel phase transformation path: solid-plasma-solid

The solid density high entropy plasma is formed in the absorption volume. The formation of the new crystalline structure *starts from a disordered state along* the paths apparently having some kinetic barriers suppressed.

The micro-explosion pathway for the synthesis of novel phases is akin to the computational method by *ab initio* random structure searching (AIRSS), (Picard and Needs, 2010, 2011).

AIRSS calculations predicted the novel silicon phases found in the micro-explosion (Rapp et al., 2015). Significant sophistication of the existing calculations is needed to model the non-equilibrium processes in hot solid density plasma.

The cooling from the plasma state to the temperature when the inter-atomic bonds begin to form via well-known heat conduction processes appears to be record fast, $\sim 10^{14}$ Kelvin/sec. The processes of the cooling down to the ambient condition and the novel phase formation during this process are complete *terra incognita*.

Until now, only the studies of the laser-transformed material were performed well after the end of the pulse. The first attempts of *in situ* measurements of the shock wave evolution during the Bessel Beam interaction with silica have been performed using the optical probe, but lacking the necessary time and space resolution (Guoyan Wang et al., 2017). Nevertheless, these experiments confirmed the propagation of the cylindrical shock waves leading to the compression of material into the cylindrical shell.

It is demonstrated experimentally that the Bessel beam transforms by 2 μJ of the laser energy almost two orders larger amounts of material, than the Gauss beam did, with the length of the focal zone of 30 μm. A further increase in the laser pulse energy allows the enlargement of the size of the focal volume and correspondingly the amount of the laser-transformed phases.

The experiments with the Bessel beam pump producing 100 μm long focal volume and a probe with the better than micron spatial resolution and better than picosecond time resolution would allow understanding the novel chaos-to-solid path for the material transformation.

The whole area modified by the Bessel beam action, including the absorption zone, the shock and heat waves affected material, is a cylinder 30–100 micron long and a few microns in diameter. The chaos-to-solid transformation occurs allegedly in an area less than micron size. The whole processes of material transformation extend from a few picoseconds to tens of nanoseconds. Thus, pump-probe time-resolved experiments are needed.

Pump-and-probe future experiments

The ideal probe, suitable for the observation of *in situ* modifications of a structure at the atomic level, is X-ray free electron laser (XFEL) with the pulse duration of 15 – 25 fs. Such lasers are operating at DESY in Hamburg (EuroXFEL, the photon energy ~8 KeV); Stanford, USA (SLAC National Acceleration Laboratory, the photon energy ~17 KeV); and Riken Institute in Japan (SACLA at Spring-8, photon energy 10 KeV, pulse duration, 25 fs, beam diameter, 0.12 nm). The combination of the ultra-short intense Bessel beam pump with the XFEL probe would be able to uncover the features of the chaos-to-solid transformations under extreme pressure. A preliminary project of such studies has been presented recently (Anand et al., 2020).

Micro-machining for 3D photonic structures

The scaling based on the laws of the mass and energy conservation allows *the prediction of the diameter of the void produced by the tightly focussed Gauss beam and diameter/length of the cylindrical channel produced by the Bessel beam*, without any fitting coefficients and in full agreement with the experiments. This scaling can be used in application of the ultra-short lasers for the precision micro-machining of transparent materials for applications in photonics.

The aim of the future study is to search the ways for increasing the energy density to the levels of a few tens MJ/cm^3 at the laboratory tabletop using the structured laser beams and to observe and control the high-pressure material transformations at the atomic space and time scale. The laser pump/ultra-short X-ray probe experiments are necessary to uncover the nature of the phase transformations and formation of novel materials.

Appendix 1

Equations, Units and Fundamental Constants in SI and Gaussian Units (Huba, 2002)

Maxwell's Equations

	Si units	Gaussian units
Faraday's law	$\nabla \times \boldsymbol{E} = -\frac{\partial \boldsymbol{B}}{\partial t}$	$\nabla \times \boldsymbol{E} = -\frac{1}{c}\frac{\partial \boldsymbol{B}}{\partial t}$
Ampere's law	$\nabla \times \boldsymbol{H} = -\frac{\partial \boldsymbol{D}}{\partial t} + \mathbf{J}$	$\nabla \times \boldsymbol{H} = -\frac{1}{c}\frac{\partial \boldsymbol{D}}{\partial t} + \frac{4\pi}{c}\mathbf{J}$
Gauss's law	$\nabla \cdot \boldsymbol{E} = \rho$	$\nabla \cdot \boldsymbol{E} = 4\pi\rho$
Gauss's law for magnetism	$\nabla \cdot \boldsymbol{B} = 0$	$\nabla \cdot \boldsymbol{B} = 0$
Material relations	$\boldsymbol{D} = \varepsilon \boldsymbol{E}$ $\boldsymbol{B} = \mu \boldsymbol{H}$	$\boldsymbol{D} = \varepsilon \boldsymbol{E}$ $\boldsymbol{B} = \mu \boldsymbol{H}$

Dimensions and Units in SI and CGSE

To get the value of a quantity in Gaussian units, multiply the value expressed in SI units by the conversion factor.

Physical parameter	Si units	Conversion factor	Gaussian units
Length; l	meter (m)	10^2	centimetre (cm)
Time; t	second (s)	1	second (s)
Velocity; v	$\frac{\text{m}}{\text{s}}$	10^2	$\frac{\text{cm}}{\text{s}}$

Physical parameter	Si units	Conversion factor	Gaussian units
Mass; m	kilogram (kg)	10^3	gram (g)
Density; ρ	$\frac{\text{kg}}{\text{m}^3}$	10^{-3}	$\frac{\text{g}}{\text{cm}^3}$
Force	newton (n)	10^5	dyne $\left(\frac{\text{gcm}}{\text{s}^2}\right)$
Energy; E	Joule (J)	10^7	erg $\left(\frac{\text{gcm}^2}{\text{s}^2}\right)$
Power; W	watt $\left(\text{W} = \frac{\text{J}}{\text{s}}\right)$	10^7	$\frac{\text{erg}}{\text{s}} = \frac{\text{gcm}^2}{\text{s}^3}$
Charge; q	coulomb	3×10^9	$\frac{\text{g}^{1/2}\text{cm}^{3/2}}{\text{s}}$
Current; i	ampere $\left(i = \frac{q}{t}\right)$	3×10^9	$\frac{\text{g}^{1/2}\text{cm}^{3/2}}{\text{s}^2}$
Electric field; **E**	$\frac{\text{volt}}{\text{m}}$	$\frac{1}{3} \times 10^{-4}$	$\frac{\text{g}^{1/2}}{\text{cm}^{1/2}\text{s}}$
Magnetic field; **H**	$\frac{\text{ampere}}{\text{M}}$	$4\pi \times 10^{-3}$	oersted $\left(\frac{\text{g}^{1/2}}{\text{cm}^{1/2}\text{s}}\right)$
Magnetic induction; **B**	tesla	10^4	gauss $\left(\frac{\text{g}^{1/2}}{\text{cm}^{1/2}\text{s}}\right)$

Fundamental Constants in SI and Gaussian Units

Physical parameter	Symbol	Value, Si units	Value, Gaussian units
Elementary charge	e	1.6022×10^{-19} C	$4.8032 \times 10^{-10} \frac{\text{g}^{1/2}\text{cm}^{3/2}}{\text{s}^2}$
Electron mass	m_e	9.1094×10^{-31} kg	9.1094×10^{-28} g
Atomic mass unit	m_a	1.6605×10^{-27} kg	1.6605×10^{-24} g

Physical parameter	Symbol	Value, Si units	Value, Gaussian units
Planck constant	h $\hbar$	$h = 6.6261 \times 10^{-34}$ J s $\hbar = h/2\pi = 1.0546 \times 10^{-34}$ Js	6.6261×10^{-27} erg s 1.0546×10^{-27} erg s
Speed of light	c	2.9979×10^{8} m/s	2.9979×10^{10} cm/s
Permittivity of free space	ε_0	8.8542×10^{-12} F/m	1
Permeability of free space	μ_0	$4\pi \times 10^{-7}$ H/m	1
Bohr radius	a_0	5.2918×10^{-11} m	5.2918×10^{-9} cm
Atomic cross section	πa_0^2	8.7974×10^{-21} m^2	8.7974×10^{-17} cm^2
Boltzmann constant	k_B	1.3807×10^{-23} J/K	1.3807×10^{-16} erg/K
Frequency associated with 1 eV		2.4180×10^{14} Hz	2.4180×10^{14} Hz
Energy associated with 1 eV		1.6022×10^{-19} J	1.6022×10^{-12} erg
Temperature associated with 1 eV		1.1604×10^{4} K	1.1604×10^{4} K
Wavelength associated with 1 eV		1.2398×10^{-6} m	1.2398×10^{-4} cm
Avogadro number	N_A	6.0221×10^{23} 1/mol	6.0221×10^{23} 1/mol

Appendix 2

Charge Oscillations (Electrons and Ions), Screening, Ideal and Non-Ideal Plasma

Cold Plasma

Thermal energy in the cold plasma is small in comparison to the Coulomb energy. The Earnshaw theorem (consequence of the Gauss theorem) states: the system of the equal number of positive and negative charges (quasi-neutral plasma) is intrinsically unstable because the uncompensated Coulomb force all the time acts at any charge. The forces continuously produce the electron density fluctuations. Equations of motion of electrons and ions under the action of fluctuating electrostatic field are as follows:

$$m_e \frac{dv_e}{dt} = e \times \Delta E \qquad \text{(A-2.1)}$$

$$M \frac{dv_i}{dt} = -e \times \Delta E$$

The fluctuating collective electrostatic field, ΔE, created by the electrons with the number density, n_e, shifts an electron to the distance, Δx. From the material equation, $\nabla \cdot E = 4\pi e n_e$, follows, $\Delta E \approx 4\pi e n_e \times \Delta x$. The equation of the electron's motion in such a field shows that electrons oscillate with the frequency, ω_{pe}:

$$\frac{d^2 \Delta x}{dt^2} = e\Delta E / m_e = \frac{4\pi e^2 n_e}{m_e} \Delta x \equiv \omega_{\text{pe}}^2 \Delta x \qquad \text{(A-2.2)}$$

Thus, any electron oscillates in quasi-neutral plasma with the so-called electron's plasma frequency:

$$\omega_{\mathrm{pe}} = \left(\frac{4\pi n_e e^2}{m_e} \right)^{1/2} \tag{A-2.3}$$

The same electrostatic field acts on the positive ion. Therefore, the ions oscillate much slower with the ion's plasma frequency:

$$\omega_{\mathrm{pi}} = \left(\frac{4\pi n_e e^2}{M} \right)^{1/2} \tag{A-2.4}$$

Non-Equilibrium, Non-Ideal Plasma

The equations of motion of the electrons and ions in the non-equilibrium plasma with the hot electrons heated by the ultra-short pulse while the ions remain cold ions are as follows:

$$m_e \frac{dv_e}{dt} = eE - k_B T_e \times \nabla \ln n_e \tag{A-2.5}$$

$$M \frac{dv_i}{dt} = -eE$$

The conventional approach neglects the electron's inertia getting the ion equation in the form

$$M \frac{dv_i}{dt} = -k_B T_e \times \nabla \ln n_e. \tag{A-2.6}$$

Ideal Plasma (Screening of the Coulomb Potential)

Plasma is in the ideal plasma state if the electrons' thermal energy is much larger than the Coulomb energy:

$$\frac{3k_B T_e}{2} \cong \frac{m_e v_e^2}{2} \gg \frac{m_e \omega_{\mathrm{pe}}^2 r_0^2}{4\pi} \tag{A-2.7}$$

Neglecting the electron's inertia in Eq. (A-2.5) and substituting the electric field with the potential, $E = -\nabla \cdot \varphi$, the electronic equation converts to

$$e \cdot \nabla \cdot \phi = \frac{k_B T_e \cdot \nabla \cdot n_e}{n_e}.$$

The electrons density reads, $n_e = n_0 \exp\{e\varphi/k_B T_e\}$. Note that $e\varphi$ is the energy of the Coulomb interaction, estimated above, $e\varphi \approx m_e \omega_{\text{pe}}^2 r_0^2$. Hence, the exponent in the expression for the electron density expresses through the Debye length, $l_{\text{De}}^2 = k_B T_e / m_e \omega_{\text{pe}}^2$, and through the average interatomic distance:

$$n_e \approx n_0 \exp\left\{\bar{r}^2 / l_{\text{De}}^2\right\}$$

In order to see the physical meaning of the Debye screening length let us come back to the coupled electrons/ions/ field equations. Coming back to the ideal plasma case ($e\varphi/k_B T_e << 1$) one can expand electron density into series over the small parameter, $n_e \approx n_0 \{1 + e\varphi/k_B T_e\}$, insert into the Poisson equation, which reduces to

$$\nabla^2 \varphi - \frac{\varphi}{l_{\text{De}}^2} = -4\pi \cdot e \cdot \delta(x).$$

The solution is well known:

$$\varphi = \frac{e}{r} \exp\{-r/l_{\text{De}}\}$$

Hence, the electrons shield out the field of a charge in the ideal plasma reducing e-fold the Coulomb field at the distance equal to the Debye length.

Coulomb Interaction in the Non-Ideal Non-Equilibrium Plasma

Let us consider the process when electrons are created during the short period of ionization, while ions remain at initial positions and temperature. The electrons have the thermal energy less than the Coulomb energy, $\varepsilon_{\text{Coul}} > k_B T_e$. Then the Debye screening of the Coulomb potential by this reason and due to the brevity of the process is negligible. The primary force coupling electrons and ions is the Coulomb force of attraction/repulsion. Our goal is to establish the time for the Coulomb force action between the swiftly created charged particles having the same charge and different masses. Let us assume the plasma is simple, i.e. the number of electrons equals to that of ions, $n_i = n_e = n$. The unscreened Coulomb force acting between charged particles in plasma having the number density n at the average interatomic distance, $r_0 = \left(\frac{3}{4\pi n}\right)^{1/3}$, is expressed as

$$F_{\text{Coul}} \approx e^2/r_0^2 = \frac{4\pi e^2 n \times r_0}{3}. \tag{A-2.8}$$

The force between electrons reads $F_{\text{Coul}}^{\text{ee}} = \frac{m_e \omega_{\text{pe}}^2 \times r_0}{3}$. Here the electron's plasma frequency is introduced, $\omega_{\text{pe}}^2 = \frac{4\pi e^2 n_e}{m_e}$. Similarly, the force acting on the ions of mass M is $F_{\text{Coul}}^{\text{ei}} = \frac{M\omega_{\text{pi}}^2 \times r_0}{3}$; here the ion plasma frequency is $\omega_{\text{pi}}^2 = \frac{m_e}{M}\omega_{\text{pe}}^2$. Therefore, the electrons start oscillating after the characteristic time $t_{C,\text{ee}} \approx \omega_{\text{pe}}^{-1} \propto (m_e)^{1/2} \approx 10^{-16}$ s, while the ions start moving after, $t_{C,\text{ei}} \approx \omega_{\text{pi}}^{-1} \propto M^{1/2}$. For example, singly ionized Silicon ions (M = 28 a.u.) start oscillating after, $t_{C,\text{ei}} \simeq 2 \times 10^{-14}$ s, while for the heavier silver plasma (M = 107.87 a.u.) this time is $t_{C,\text{ei}} \simeq 3.9 \times 10^{-14}$ s. Hence the coupling of electrons and ions via the Coulomb field occurs much faster than the energy transfer via the collisions. Note that the energy of the Coulomb attraction (height of the barrier for electron preventing to leave plasma), is $\varepsilon_{\text{Coul}} \approx e^2/r_0$ (taking $r_o = 1.33 \times 10^{-8}$ cm, $n \approx 10^{23}$ cm^{-3}).

If the intense laser with the pulse shorter than $t_{C,\text{ei}}$ produces free electrons with the thermal energy in excess of the work function, the electrons escape from plasma having no time to couple to ions by the attraction force. The ions stripped off electrons might be ejected from solid plasma by the force of the Coulomb repulsion if the enough electrons leave a solid.

Electron and Ion Plasma Frequency in the Ionized Sapphire

The intense laser converts sapphire into the solid density plasma (the average ion mass, M_{av} = 20 a.u.; $n_e \approx n_i \approx 10^{23}$ cm^{-3}), where $\omega_{\text{pe}} = 1.78 \times 10^{16}$ s^{-1}, and $\omega_{\text{pi}} = 0.93 \times 10^{14}$ s^{-1}. For example in sapphire these times are 0.056 fs and 11 fs.

The electric field of the laser beam at intensity 10^{14} W/cm^2 equals to, $E = (8\pi I/c)^{1/2} \approx 0.9 \times 10^6$ CGSE. Electron's oscillation energy in such a field is $\varepsilon_{\text{osc}} \sim 9.6$ eV (at the wavelength equal to micron, μm).

The electron-ion energy transfer time is

$$t_{\text{energy}} = \left(\frac{m_e}{M_{\text{av}}} v_{\text{ei}} \right)^{-1} = 7.3 \text{ ps.}$$

Non-equilibrium plasma created in a dielectric by the swift action of the ultra-short laser pulse has the additional feature in comparison to the simple plasma considered above. A transparent dielectric converts to the absorbing plasma when the real part of the permittivity turns to zero. It means that the negative polarization from the newly created electrons compensates the positive polarization of the remaining neutrals. At this stage, the number of electrons is much less the number of neutral atoms. Therefore one may suggest that the fluctuation of the electrostatic field in the newly created plasma, which is intermixed with the partly ionized dielectric, is enhanced by the presence of dielectric, $\varepsilon_0 \Delta E \approx 4\pi e n_e \times \Delta x$, resulting in the decrease of the plasma frequency, $\omega_{\text{pe}}^2 = \frac{4\pi e^2 n_e}{\varepsilon_0 m_e}$. Obviously at the full first ionization ($n_e \approx n_a$) the plasma frequency retains its conventional form.

Note that inverse electron plasma frequency is comparable to the typical timescale for the inter-atomic processes, $a_B/v_a \sim 10^{-16}$ s. This is also a timescale of the tunnelling ionization of an atom under the action of the intense ultra-short pulse, i.e. time for plasma creation. The collective mode of the Coulomb interactions in plasma between electrons and ions establishes during much longer time, $t_{\text{ei}} = \omega_{\text{pi}}^{-1} = M\omega_{\text{pe}}^{-1}/m_e$.

Appendix 3

Melting Parameters for Metals and Dielectrics in Equilibrium

ε_b, binding (cohesive) energy (in eV); T_m, melting point (in eV and in Kelvin); ΔS_f, the entropy of fusion (melting) in units of the Boltzmann constant; ΔH_f, the enthalpy of fusion in kJ/cm^3.

	ε_b(eV)	T_m	T_m/ε_b	ΔH_f	ΔS_f	Comments
Al	3.39	0.08 eV (928.5 K°)	0.0236	1.07	1.378 k_B	$n_{liquid} < n_{solid}$
Cu	3.49	0.117 eV (1,357.9 K°)	0.0335	1.865	1.17 k_B	"
Ag	2.95	0.1065 eV (1,236 K°)	0.0361	1.097	1.098 k_B	"
Fe	4.28	0.156 eV (1,810 K°)	0.0364	1.95	1.296 k_B	"
Au	3.81	0.1154 eV (1,139.3 K°)	0.0303	1.23	1.13 k_B	"
Bi	2.16	0.047 (545.5 K°)	0.0217	0.5	2.4 k_B	$n_{liquid} > n_{solid}$
Si	4.63	0.1454 (1,687.5 K°)	0.0314	4.17	3.725 k_B	"
Ga	2.81	0.026 (301.7 K°)	0.0093	0.474	2.267 k_B	"
InSb	2.65	0.0685 (795 K°)	0.0258	0.975	3 k_B	"

Appendix 4

Effective Collision Rates for Al, Cu and Ag Extracted from the Optical Measurements at the Room Temperature, 295 K

The effective collision rate under assumption that the dielectric permittivity obeys the Drude form expresses through the real and imaginary parts of the permittivity, measured at the different wavelength at room temperature (Palik, 1998), as follows: $\nu_{\text{eff}} = \varepsilon_{\text{im}} \cdot \omega/(1 - \varepsilon_{\text{re}})$.

Aluminium

λ (μm)	$\omega \times 10^{15}$ (s^{-1})	ε_{re}	ε_{im}	$\nu_{\text{eff}} \times 10^{15}$ (s^{-1})
0.5	3.77	−36.375	9.35	0.94
0.6	3.14	−51.27	17.42	1.046
0.799	2.356	−(63.56÷ 56.04)	47.32÷ 49.13	2.02
0.900	2.094	−64.65	34.2	1.09
1.000	1.885	−89.95	25.87	0.53
1.520	1.26	−253.3	42.5	0.156

Copper

λ (μm)	$\omega \times 10^{15}$ (s^{-1})	ε_{re}	ε_{im}	$\nu_{\text{eff}} \times 10^{15}$ (s^{-1})
0.4133	4.56	−3.49	5.21	5.297
0.5166	3.65	−5.5	5.824	3.27
0.6199	3.04	−10.42	1.76	0.469
0.7084	2.66	−17.93	1.814	0.255
0.8265	2.28	−27.6	2.735	0.218

Silver

λ (μm)	$\omega \times 10^{15}$ (s^{-1})	ε_{re}	ε_{im}	$\nu_{eff} \times 10^{15}$ (s^{-1})
0.4959	3.8	−8.277	0.7488	0.306
0.5904	3.19	−13.38	0.886	0.195
0.6888	2.74	−19.694	1.243	0.1646
0.8266	2.28	−30.23	1.595	0.115
1.033	1.82	−48.81	3.159	0.115
1.24	1.52	−71.97	5.586	0.116

Appendix 5

Material and Optical Properties of Metals Necessary for the Calculation of the Metal Ablation by the Ultra-Short Laser Pulses

Table A-5.1 Physical properties of metals, Al, Cu, Fe and Pb

	Al	Cu	Fe	Pb
Thickness of mono-layer, 10^{-8} cm	2.86	2.56	2.35	3.5
Electron density, 10^{22} cm^{-3}	18.6	8.45	16.8	13.2
Atomic density, 10^{22} cm^{-3}	6.02	8.45	8.5	3.3
Fermi energy, eV	11.63	7.0	11.1	9.47
Ionisation potential, eV	5.86	7.73	7.9	7.417
Binding energy, eV	3.065	3.173	3.695	1.795
Thermal diffusivity, cm^2/s	0.979	1.165	0.228	0.241
m^*/m_e (Kittel, 1996)	1.48	1.38	8	1.97
$\omega_{pe} \times 10^{16}$ s^{-1}	2	1.4	0.817	1.46
$(m_e^*/M) \times 10^{-5}$	3	1.19	7.85	0.52
Ion's mass, M (atomic units)	26.9858	63.546	55.845	207.2

Note: $\omega_{pe} = (4\pi e^2 n_e/m_e^*)^{1/2}$, plasma frequency for the electrons in the conduction band; m^*/m_e, ratio of the electron effective mass to the free electron mass.

Table A-5.2 The optical parameters of Al, Cu, Fe, for λ = 532 nm, and Pb (λ = 589.3 nm) at the room temperature

	Al	Cu	Fe	Pb*
n	0.85	2.60	1.05	2.01
κ	6.48	2.58	3.33	3.48
A	0.075	0.487	0.432	0.38
l_s, (nm)	13.11	32.85	25.43	24.3
t^{en}_{e-ph}, (ps)	1.67	6	1.56	13

*At 589.3 nm.

Note: n and κ are the real and imaginary parts of the refractive index; $A = {4n\kappa}/{(n+1)^2+k^2}$ is the Fresnel absorption coefficient; $l_s = c/\omega\kappa$, absorption depth; $\nu^{en}_{e-ph} \approx \omega_{pe}(m^*_e/M)$ energy exchange rate; and $t^{en}_{e-ph} = (\nu^{en}_{e-ph})^{-1}$ energy exchange time.

The electron heat capacity changes from that for the degenerated electron gas at low temperature to that for the ideal gas at high temperature. Interpolation, $C_e \approx \frac{3}{2}k_B(2x - x^2)$ where, $x = k_BT_e/\varepsilon_F$, attains the ideal gas value at $x = 1$.

Table A-5.2 The electron heat capacity at $k_BT_e = \varepsilon_b$ (supposedly near the ablation threshold)

	Al	Cu	Fe	Pb
Binding energy, eV	3.065	3.173	3.695	1.795
C_e (in units k_B)	0.473	0.76	0.122	0.269

Appendix 6

Spatial Separation of Ions with Different Masses during the Electron-to-Ion Energy Transfer in the Non-Equilibrium Multi-Component Plasma ($T_e > T_i$)

The rate of electron–ion energy transfer rate is inversely proportional to the ion's mass, $\nu_{ei}^{en} = \frac{m_e \times \nu_{ei}^{mom}}{M_i}$. Therefore, in plasma containing ions with the different masses the light ions get their temperature equilibrated with that of electrons faster than the heavy ions, $t_L = (\nu_{ei}^{en})^{-1} = \frac{M_L}{m_e \times \nu_{ei}^{mom}}$. During the period $(t_H - t_L)$, the light ions diffuse away from the place where they were close to the heavy ions. For example, during the ultra-fast ionization of sapphire (Al_2O_3) or silica (SiO_2) oxygen ions remain close to the heavy ones until the energy transfer from the heated electrons. The diffusion velocity of the light ion is $D_L = l_{mfp} \times v_L/3$. Therefore, the light ion departs from the heavy counterpart to the distance $L_{sep} = [D_L \times (t_H - t_L)]^{1/2}$. Now the heavy-light ions separation distance can be presented explicitly:

$$L_{sep} = \left[\frac{l_{mfp} \times v_L \times (M_H - M_L)}{3m_e \times \nu_{ei}^{mom}}\right]^{1/2}$$

Example: Separation of Si and oxygen ions in the ionized silica at $T_e \sim 100$ eV; $M_{Si} = 28$ a.u.; $M_O = 16$ a.u.; v_O (100 eV) = 0.425×10^7 cm/s; mean free path (inter-atomic distance), $l_{mfp} = 2.5 \times 10^{-8}$ cm; ν_{ei}^{mom}(100 eV) $\approx 2 \times 10^{14}$ s^{-1} (plasma is non-ideal). Thus, Si–O separation distance is $L_{Si\text{-}O} \approx 20$–$30$ nm (approximately 10 inter-atomic distances).

Appendix 7

Material and Optical Properties of Dielectrics (SiO_2, Si, Al_2O_3) at the Room Temperature

n_a, atomic number density; ρ_0, mass density (g/cm^3); Y, Young modulus (GPa); $p_{cold} = \rho_0 v^2_{sound}$, cold pressure (GPa); v_{sound}, speed of sound (cm/s); Δ_g band gap (eV); D, thermal diffusion coefficient (cm^2/s); ε_b, binding energy per atom (eV).

	n_a 10^{22} cm^{-3}	ρ_0 (g/cm^3)	Y (GPa)	$\rho_0 v^2_{sound}$ (GPa)	v_{sound} cm/s	Δ_g (eV)	D (cm^2/s)	ε_b (eV)
SiO_2	6.6	2.196	78	54.9	5×10^5	~ 7.6	0.02	3.1
Si	7	2.329	130–188	168	8.5×10^5	1.12	0.0006	3.97
Al_2O_3	12	3.95–4.1	400	~ 310		7.3–9.6	0.06–0.084	5.5

Dielectric Permittivity for SiO_2, Si and Al_2O_3

For the light with the wavelength, λ = 800 nm (photon's energy, $\hbar\omega$ = 1.55 eV; $\omega_l = 2.356 \times 10^{15}$ s^{-1}; critical electrons' density $n_c = 1.74 \times 10^{21}$ cm^{-3})

	ε_0
SiO_2	2.11
Si	13.7
Al_2O_3	3.13*

*Dobrovinskaya et al. (2009).

Silicon: The first and second ionization potential, J_1 = 8.16 eV; J_2 = 16.3 eV.

Sapphire: Effective electron collision rates, at the room temperature (electron–phonon rate) $\nu_{e\text{-}ph}(T_{room}) = k_B T_{room}/\hbar = 3.8 \times 10^{13}\ s^{-1}$; maximum rate (electron–ion near the ablation threshold), $\nu_{max} = 5 \times 10^{15}\ s^{-1}$; ionization potential, $J_{Al} = 6$ eV; $J_O = 13.6$ eV; average ion, $J_{Al\text{-}O} = 10$ eV; melting point, 2,345 K; boiling point, 3,250 K.

Appendix 8

Ionization Time and Optimum Intensity for the Tunnel Ionization

The Time an Electron Needs to Move from the Valence Band to the Conduction Band under the Action of Short Intense Pulse (Keldysh, 1965)

The width of the potential barrier in a dielectric with the band gap, Δ_g, in the electric field, E is as follows:

$$\Delta l_b = \frac{\Delta_g}{eE}$$

The electron oscillating in the high frequency electric field with velocity, $v_{os} = eE/m\omega$, moves through the barrier during the time

$$t_b = \frac{\Delta l_b}{v_{os}} = \frac{\Delta_g}{\varepsilon_{os}} \times \frac{1}{2\omega} \propto \frac{c}{\lambda}.$$

Here $\varepsilon_{os} = \frac{mv_{os}^2}{2} = 9.35\ \text{eV}\left(\frac{\text{I}}{10^{14}\ \text{W/cm}^2}\right)\left(\frac{\lambda}{\mu\text{m}}\right)^2$ is the electron's oscillation energy. Ionization time is now the explicit function of the laser intensity and wavelength:

$$t_b = \left(\frac{\Delta_g}{9.35\ \text{eV}}\right)\left(\frac{\text{I}}{10^{14}\ \text{W/cm}^2}\right)^{-1}\left(\frac{\lambda}{\mu\text{m}}\right)^{-1} 2.65 \times 10^{-16}\ \text{s}$$

The above time is of the typical scale for the inter-atomic processes, a_B/v_a, and also a time for the Coulomb interaction of two charges. One can see that the wide band gap dielectrics (such a glass and sapphire) in the intense field (>10^{14} W/cm^2) might be efficiently ionized in the tunnelling regime even by very short (few fs) pulses.

Classical Picture of Ionization in a Stationary External Field (Bethe and Salpeter, 1977; Gibbon, 2005)

The Coulomb potential of electron in the field of eZ-charged ion is modified by the external homogeneous electric field, E, as follows:

$$V(x) = -\frac{Ze^2}{x} - eEx$$

At some distance, x_m, from the ion, the external field suppresses the Coulomb potential below the ionization energy, J_i, letting an electron to escape spontaneously (over-the-barrier-tunnelling = OBT). From the condition, $\frac{dV(x_m)}{dx} = 0$, one finds $x_m^2 = Ze/E$. Now the amplitude of the critical electric field, E_c, capable of suppressing the Coulomb potential below the ionization energy follows from the condition, $V(x_m) = J_i$:

$$E_c = J_i^2/4Ze^3$$

Considering the above field as the peak field of the linearly polarized laser, one expresses the critical intensity, $I_c = cE_c^2/8\pi$ in the following form:

$$I_c = cJ_i^4/128\pi Z^2 e^6$$

It is instructive to express the above formula through the fundamental constants taking $Z = 1$, $J_i = \frac{e^2}{2a_B} = 13.61$ eV. Then one obtains $I_c = \frac{I_{\text{at}}}{256} \approx 1.4 \times 10^{14}$ W/cm^2.

Appendix 9

Time-Dependent Electrons Number Density Produced by the Only MPI Mechanism—Comparison to the Impact Ionization (Analytics)

The equation for the electrons number density, n_e, produced by the only MPI (saturation on the full first ionization) reads

$$\frac{dn_e}{dt} = (n_a - n_e) \cdot w_{\text{MPI}}.$$

We approximate the MPI ionization rate, w_{MPI}, by the power function where m stands for the ratio of the band gap to the photon's energy, $m = \Delta_g/\hbar\omega$:

$$w_{\text{MPI}} = \beta_m \left[\frac{I(t)}{I_0} \right]^m \ [\text{s}^{-1}]$$

Time dependence of the laser intensity has the Gauss form:

$$I(t) = I_m \cdot f(t) = I_m \cdot \exp\{-4(1 - 2t/t_p)^2\}$$

The rate equation reduces to the dimensionless form introducing the variables $\delta_e = \dfrac{n_e}{n_a}$; $\theta = t/t_p$:

$$\frac{d\delta_e}{d\theta} = (1 - \delta_e) \times A_m(\theta)$$

Here $A_m(\theta) = B_m \cdot \exp\{-4m(1 - 2\theta)^2\}$. The combination of the laser and material parameters reads $B_m = t_p \cdot \beta_m \cdot (I_{\max}/I_0)^m$. The integration produces the ionization degree, n_e/n_a, as a function of time, laser and material parameters:

$$\delta_e(t/t_p) = 1 - \exp\left\{ \int_0^{t/t_p} A_m(\theta) \cdot d\theta \right\}$$

The integral is expressed through the error function:

$$\int_0^{t/t_p} A_m(\theta) \cdot d\theta = \frac{B_m}{4m^{1/2}} \cdot \left\{ - \int_{2m^{1/2}}^{-x} \exp\{-x^2\} dx \right\}$$

$$= \frac{B_m \sqrt{\pi}}{8m^{1/2}} \cdot [\mathrm{erf}(2m^{1/2}) - \mathrm{erf}(2m^{1/2}(1 - 2t/t_p))]$$

Taking into account that $m > 1$ and erf(2) = 1, the ionization degree reads

$$\frac{n_e(t/t_p)}{n_a} = 1 - \exp\left\{ - \frac{B_m \sqrt{\pi}}{8m^{1/2}} \cdot \left[1 - \mathrm{erf}(2m^{1/2}(1 - 2t/t_p))\right] \right\}.$$

The ionization degree to the end of the pulse is

$$\frac{n_e(t/t_p)}{n_a} = 1 - \exp\left\{ - \frac{\sqrt{\pi}}{4m^{1/2}} \cdot t_p \cdot \beta_m \cdot (I_0/I_1)^m \right\}.$$

The laser fluence reads

$$F(t) = I_0 \cdot t_p \cdot \int_0^{\theta} \exp\{-4(1 - 2\theta)^2\} \cdot d\theta$$

$$= \frac{I_0 \cdot t_p \cdot \sqrt{\pi}}{8} (1 - \mathrm{erf}(1 - 2t/t_p)).$$

The fluence at the end of the pulse follows:

$$F(t_p) = \frac{I_0 \cdot t_p \cdot \sqrt{\pi}}{8} (1 + \mathrm{erf}(1)) = 0.408 \cdot I_{\max} \cdot t_p; \ \mathrm{erf}\ (1) = 0.8427$$

The ionization degree reads

$$\frac{n_e(t/t_p)}{n_a} = 1 - \exp\left\{-\frac{B\sqrt{\pi}}{8m^{1/2}} \cdot [1 - \mathrm{erf}(2m^{1/2}(1 - 2t/t_p))]\right\}.$$

Thus, one gets the ionization degree to the end of the pulse

$$\frac{n_e(t_p)}{n_a} = 1 - \exp\left\{-\frac{\sqrt{\pi}}{4m^{1/2}} \cdot t_p \cdot \beta_m \cdot (I_{\max}/I_0)^m\right\}.$$

At small ionization degree $\frac{\sqrt{\pi}}{4m^{1/2}} \cdot t_p \cdot \beta_m \cdot (I_{\max}/I_0)^m << 1$, one can expand the above in the series:

$$\frac{n_e(t/t_p)}{n_a} \approx \frac{\sqrt{\pi}}{4m^{1/2}} \cdot t_p \cdot \beta_m \cdot (I_0/I_1)^m$$

Example: Laser parameters necessary for converting the fused silica into plasma, comparison of the MPI and the impact ionization rates

Let us estimate with the above formulae the pulse duration and maximum intensity necessary for the conversion of silica into plasma. The number density of electrons necessary for the silica-to-plasma conversion follows from the condition when the real part of the permittivity turns zero, $\varepsilon_{re} = 0$. For silica at 800 nm, it equals $n_e^{thr} \approx \varepsilon_0 \cdot n_{cr} = 3.7 \times 10^{21}$ cm^{-3} giving $\delta_e = 0.056$ ($n_a = 6.6 \times 10^{22}$ cm^{-3}). The multi-photon rate for the fused silica with the band gap $\Delta_g = 7.6$ eV takes a form of 5-photon MPI (800 nm; 1.55 eV; $m \approx 5$):

$$w_{\mathrm{mpi}} = 1.3 \times 10^{10} (I/10^{13}\,\mathrm{W/cm^2})^5 \mathrm{s}^{-1}.$$

The ionization rate by the electrons impact reads

$$w_{\mathrm{imp}} = \alpha I(\mathrm{s}^{-1});\ \alpha = 10\ \mathrm{cm^2/s \times W}$$

$$\frac{n_e(t_p)}{n_a} = 1 - \exp\left\{-0.198 \cdot t_p \cdot \beta_5 \cdot (I_{\max}/10^{13}\,\mathrm{W/cm^2})^5\right\}$$

From the above formulae follows that such ionization degree solely by the MPI mechanism can be achieved by 100 fs long Gauss pulse with the maximum intensity of $I_{max} = 2.94 \times 10^{13}$ W/cm^2 at the fluence 1.2 J/cm^2. The ionization rate by the MPI at this intensity constitutes $n_a \times w_{MPI} = 6.6 \times 10^{22} \times 2.88 \times 10^{12}$ cm^{-3} s^{-1} = 1.9×10^{35} cm^{-3} s^{-1}, while the impact rate is $n_e \times w_{imp} = 3.7 \times 10^{21}$ cm^{-3} $\times$ 2.94×10^{14} cm^{-3} s^{-1} = 1.09×10^{36} cm^{-3} s^{-1}. Therefore, the contribution of the impact ionization becomes dominant before conversion into plasma significantly reducing the ionization threshold fluence below one J/cm^2.

Bibliography

Abell, G. C. (1985), Empirical chemical pseudopotential theory of molecular and metallic bonding, *Phys. Rev. B*, 31, 6184–6196.

Afanasiev, Y. V., N. N. Demchenko, I. N. Zavestovskaya, V. A. Isakov, A. P. Kanavin, S. A. Uriupin, and B. N. Chichkov (1999), Simulation of metal ablation by the ultra short laser pulses, *Izvestya Acad. Nauk Seria Fiz.*, 63, 677 (in Russian).

Afanasiev, Y. V., and O. N. Krokhin (1971), High temperature and plasma phenomena arising in the power laser-matter interaction, in: *Physics of High Energy Density*, ed. P. Calderola and H. Knoepfel (Academic Press).

Ahrens, T. (1980), Dynamic compression of earth materials, *Science*, 207, 1035–1041.

Akhmanov, S. A., V. A. Vyspoukh, and A. S. Chirkin (1988), *Optics of Femtosecond Laser Pulses*, Nauka (in Russian).

Allen, P. B. (1987), Theory of thermal relaxation of electrons in metals, *Phys. Rev. Lett.*, 59, 1460–1463.

Allen, P. B. (2000), Misbehavior of metals. *Nature*. 405, 1007.

Ammosov, M. V., N. B. Delone, and V. P. Krainov (1986), Tunnel ionization of complex atoms and of atomic ions in an alternating electromagnetic field, *Sov. Phys. JETP*, 64(6), 1191–1194.

Anisimov, S. I., Y. A. Imas, G. S. Romanov, Y. V. Khodyko (1971), *Action of High Intensity Radiation on Metals*, National Technical Information Service, Springfield, Virginia.

Anisimov, S. I., B. L. Kapeliovich, and T. L. Perel'man (1974), Electron emission from metal surfaces exposed to ultrashort laser pulses, *Sov. Phys. JETP*, 39(2), 375–377.

Anisimov, S. I., and B. S. Luk'yanchuk (2002), Selected problems of laser ablation theory, *Phys. Uspekhi*, 45(3), 293–324.

Avrorin, E., V. Simonenko, L. Shibarshov (2006), Physical studies with nuclear explosions, *Phys. Uspekhi*, 174, 449.

Balling, P., and J. Schou (2013), Femtosecond-laser ablation dynamics of dielectrics: Basics and applications for thin film, *Rep. Prog. Phys.*, 76, 036502 (39 pp.).

Bethe, H. A., and E. E. Salpeter (1977), *Quantum Mechanics of One- and Two- Electron Atoms. Plenum*, New York.

Beuton, R., B. Chimier, J. Breil, D. Hébert, P.-H. Maire, and G. Duchateau (2017), Thermo-elasto-plastic simulations of femtosecond laser-induced structural modifications: Application to cavity formation in fused silica. *J. Appl. Phys.*, 122, 203104; https://doi.org/10.1063/1.4993707.

Bomzon, Z., M. Gu, and J. Shamir (2006), Angular momentum and geometrical phases in tight-focused circularly polarized plane waves, *App. Phys. Lett.*, 89, 241104.

Born, M. (1939), Thermodynamics of crystals and melting, *J. Chem. Phys.*, 7, 591–603.

Born, M., and E. Wolf (2003), *Principles of Optics*, Cambridge.

Boschetto, D., E. G. Gamaly, A. V. Rode, B. Luther-Davies, D. Glijer, T. Garl, O. Albert, A. Rousse, and J. Etchepare (2008), Small atomic displacements recorded in bismuth by the optical reflectivity of femtosecond laser-pulse excitations, *Phys. Rev. Lett.*, 100, 027404.

Boschetto, D., T. Garl, and A. Rousse (2010), Ultrafast dielectric function dynamics in bismuth, *J. Modern Opt.*, 57, 953–958.

Cahn, R. W. (1986), Melting and the surface, *Nature*, 323, 668–669.

Carr, C. W., Radousky H. B., and S. G. Demos (2003), Wavelength dependence of laser-induced damage: Determining the damage initiation mechanisms, *Phys. Rev. Lett.*, 91, 127402.

Carr, C. W., H. B. Radousky, A. M. Rubenchik, M. D. Feit, and S. G. Demos (2004), Localized dynamics during laser-induced damage in optical materials, *Phys, Rev. Lett.*, 92, 087401-1.

Cauble, R., D. W. Phillion, T. J., Hoover, N. C. Holmes, J. D. Kilkenny, and R. W. Lee (1993), Demonstration of 0.75 Gbar planar shocks in x-ray driven colliding foils, *Phys. Rev. Lett.*, 70, 2102.

Cauble, R., and W. Rozmus (1995), Two-temperature frequency dependent electrical resistivity in solid density plasma produced by ultra-short pulses, *Phys. Rev. E*, 52, 2974–2981.

Cheng, T. K., Vidal, J., Zeiger, H. J., Dresselhaus, G., Dresselhaus, M. S., and Ippen, E. P. (1991), Mechanism for displacive excitations of coherent phonons in Sb,Bi, Te, and Ti_2O_3, *Appl. Phys. Lett.*, 59, 1923–1925.

Chernyi, G. G. (1957), The problem of a point explosion, *Dokl. Acad. Nauk SSSR*, 112, 213–216, Transl. as Rep. No.MOA TIL/T.4871, Ministry of Aviation (Gt. Brit.) 1959.

Chimier, B., O. Utéza, N. Sanner, M. Sentis, T. Itina, P. Lassonde, F. Légaré, F. Vidal, and J. C. Kieffer (2011), Damage and ablation thresholds of fused-silica in femtosecond regime, *PRB*, 84, 094104.

Comins, N. R. (1972), The optical properties of liquid metals, *Phil. Mag.*, 25, 817.

Courvoisier, F., J. Zhang, M. K. Bhuyan, M. Jacquot, and J. M. Dudley (2012), Applications of femtosecond Bessel beams to laser ablation, *Appl. Phys. A*, 112, 29–34.

Denisov, N. G. (1957), On a singularity of the field of an electro-magnetic wave propagated in an inhomogeneous plasma. *Sov. Phys. JETP.*, 4, 544.

Dobrovinskaya, E. R., L. A. Lytvynov, and V. Pishchik (2009), *Sapphire: Material, Manufacturing, Applications,* Springer.

Du, D., X. Liu, G. Korn, J. Squier, and G. Mourou (1994), Laser-induced break down by impact ionisation in SiO_2 with pulse width from 7 ns to 150 fs, *Appl. Phys. Lett.*, 64(23), 3071–3073.

Durnin, J., J. J. Miceli, and J. H. Eberly (1987a), Non-diffracting light beams, *Phys. Rev. Lett.*, 58, 1499.

Durnin, J., (1987b) Exact solutions for nondiffracting beams. I. The scalar theory. *J. Opt. Soc. Am.*, A, 4, 651–654.

Eidmann, K., J. Meyer-ter-Vehn, T. Schlegel, and S. Huller (2000), Hydrodynamic simulation of subpicosecond laser interaction with solid density matter, *Phys. Rev. E*, 62, 1202.

Fecht, H. J. (1992), Defect-induced melting and solid-state amorphization, *Nature*, 356, 133–135.

Fecht, H. J., and W. L. Johnson (1988), Entropy and enthalpy catastrophe as a stability limit for crystalline material, *Nature*, 334, 50–55.

Fortov, V. (2005), *Intense Shock Waves and Extreme States of Matter*, Bukos.

Forsterling, K. (1949), The Propagation, with Oblique Incidence, of Elec-tromagnetic Waves in a Stratified Medium under the Influence of a Magnetic Field. In: *Archiv der elektrischen Übertragung*. Bd. 3. S. 115–120.

Fradin, D. W., N. Bloembergen, and J. P. Letellier (1973), Dependence of laser-induced breakdown field strength on pulse duration, *Appl. Phys. Lett.*, 22, 635–637.

Friedli, C., and Ashcroft, N. W. (1975), Aluminum under high pressure. I. Equation of state, *Phys. Rev. B*, 12, 5552–5559.

Gamaly, E. (2010), Ultra-fast disordering by fs-lasers: Lattice superheating prior to the entropy catastrophe, *Appl. Phys. A*, 101, 205–208.

Gamaly, E. G. (2011), *Femtosecond Laser-Matter Interactions: Theory, Experiments and Applications*, Pan Stanford Publishing.

Gamaly, E., S. Joudkazis, H. Misawa, B. Luther-Davies L. Hallo, P. Nicolai, and V. T. Tikhonchuk (2006), Laser-matter interaction in a bulk of a transparent solid: Confined micro-explosion and void formation, *Phys. Rev. B*, 73, 214101.

Gamaly, E. G., S. Juodkazis, and A. V. Rode (2018), Extreme energy density confined inside a transparent crystal: Status and perspectives of solid-plasma-solid transformations, *Nanomaterials*, 8, 555; doi:10.3390/nano8070555.

Gamaly, E. G., N. R. Madsen, M. Duering, A. V. Rode, V. Z. Kolev, and B. Luther-Davies (2005), Ablation of metals with picosecond laser pulses: Evidence of long-lived non-equilibrium conditions at the surface, *Phys. Rev. B*, 71, 174405.

Gamaly, E. G., L. Rapp, V. Roppo, S. Juodkazis, and A. V. Rode (2013c), Generation of high energy density by fs-laser induced confined micro-explosion, *New J. Phys.*, 15, 025018.

Gamaly E. G., and A. V. Rode (2013a), Physics of ultra-short laser interaction with matter: From phonon excitation to ultimate transformations, *Prog. Quantum Electronics*, 37, 215–323.

Gamaly, E. G., and A. V. Rode (2013b), Ultra-fast electronic relaxation in super-heated bismuth, *New J. Phys.*, 15, 013035.

Gamaly, E. G., and A. V. Rode (2014), Transient optical properties of dielectrics and semiconductors excited by an ultra-short laser pulse, *JOSA B*, 31, C36–C43.

Gamaly, E. G., and A. V. Rode (2016), [INVITED] Coupling of polarization of high frequency electric field and electronic heat conduction in laser created plasma, *Opt. Laser Technol.*, 82, 69–71.

Gamaly, E. G., and A. V. Rode (2018), Ultrafast re-structuring of the electronic landscape of transparent dielectrics: New material states (Die-Met), *Appl. Phys. A*, 124, 278; https://doi.org/10.1007/s00339-018-1693-3.

Gamaly, E. G., A. V. Rode, and B. Luther-Davies (1999), Ultrafast ablation with high-pulse-rate lasers. Part I: Theoretical considerations, *J. Appl. Phys.*, 85(8), 4213–4221.

Gamaly, E. G., A. V. Rode, and B. Luther-Davies (2007), Ultra-fast laser ablation and film deposition, in: *Pulsed Laser Deposition of Thin Films: Applications in Electronics, Sensors, and Biomaterials,* ed. R. W. Eason (John Wiley & Sons Inc), pp. 99–130.

Gamaly, E. G., A. V. Rode, B. Luther-Davies, and V. T. Tikhonchuk (2002), Ablation of solids by femtosecond lasers: Ablation mechanism and ablation thresholds for metals and dielectrics, *Phys. Plasmas,* 9, 949–957.

Gamaly, E. G., A. V. Rode, L. Rapp, R. Giust, L. Furfaro, P. A. Lacourt, J. M. Dudley, F. Courvoisier, and S. Juodkazis (2017), Interaction of the ultra-short Bessel beam with transparent dielectrics: Evidence of high-energy concentration and multi-TPa pressure, arXiv:1708.07630, 28 August 2017.

Gamaly, E. G., A. V. Rode, O. Uteza, V. Kolev, B. Luther-Davies, T. Bauer, J. Koch, F. Korte, and B. N. Chichkov (2004), Control over a phase state of the laser plume ablated by femtosecond laser: Spatial pulse shaping, *J. Appl. Phys.*, 95, 2250–2257.

Garl, T. (2008), PhD thesis: *Ultrafast Dynamics of Coherent Optical Phonons in Bismuth*, Ecole Polytechnique, Palaiseau, France (unpublished).

Garl, T., E. G. Gamaly, D. Boschetto, A. V. Rode, B. Luther-Davies, and A. Rousse (2008), Birth and decay of coherent optical phonons in femtosecond-laser-excited bismuth, *Phys. Rev. B*, 78, 134302.

Gallagher, T. F. (1992), Rydberg Atoms in Strong Microwave Fields. In: *Atoms in Intense laser fields*. Ed. M. Gavrila (Academic Press, Boston) 67–108.

Genieys, T., M. Sentis, and O. Utéza (2019), Investigation of ultra-short laser excitation of aluminum and tungsten by reflectivity measurements, *Appl. Phys. A*, 126(4), 263; doi: 10.1007/s00339-020-3440-9.

Gibbon, P. (2005), *Short Pulse Laser Interactions with Matter*, Imperial College Press.

Ginzburg, V. L. (1964), *The properties of Electromagnetic Waves in Plasma*, Pergamon.

Gilden'burg, V. B. (1964), Resonance properties of inhomogeneous plasma structures. *Sov. Phys. JETP.*, 18, 1359.

Glezer, E., and Mazur, E. (1997), Ultrafast-laser driven micro-explosions in transparent materials, *Appl. Phys. Lett.*, 71, 882–884.

González de Alaiza Martínez, P., G. Duchateau, B. Chimier, and R. Nuter, I. Thiele, S. Skupin, and V. T. Tikhonchuk (2018), Maxwell-consistent, symmetry- and energy-preserving solutions for ultrashort-laser-pulse propagation beyond the paraxial approximation, *Phys. Rev. A*, 98, 043849.

Gorecki T. (1977), Comments on vacancies and melting, *Scripta Metallurgica*, 11, 1051–1053.

Gray, D. E., ed. (1972), *American Institute of Physics Handbook*, 3rd edition, McGraw-Hill Book Company.

Grimvall, G. (1981), *The Electron-Phonon Interaction in Metals, Selected Topics in Solid State Physics*, Vol. XVI, ed. E. P. Wohlfarth, North-Holland.

Guo, C., Rodriguez, G., A. Lobad, and A. J. Taylor (2000), Structural phase transition of aluminium induced by electron excitation, *Phys. Rev. Lett.*, 84, 4493–4496.

Hicks, D. G., P. M. Celliers, G. W. Collins, J. H. Eggert, and S. J. Moon (2003), Shock-induced transformation of Al_2O_3 and LiF into semiconducting liquids, *Phys. Rev. Lett.*, 91, 035502.

Hodgson, J. (1962), Optical properties of liquid indium, cadmium, bismuth and antimony, *Phil. Mag.*, 7, 229–236.

Holway, L. H. Jr, and D. W. Fradin (1975), Electron avalanche breakdown by laser radiation in insulating crystals, *J. Appl. Phys.*, 46(1), 279–291.

Huba, J. D. (2002), *NRL Plasma Formulary*, Naval Research Laboratory, Washington DC.

Il'inskii, Y. A., and L. V. Keldysh (1994), *Electromagnetic response of Material Media*, Plenum.

Iacquemot, S., and Decoster, A., (1991), Z scaling of collisional Ne-like X-ray lasers using exploding foils: refraction effects, *Laser Part. Beams*, 9, 517.

Jarutis, V., R. Paskauskas, and A. Stabinis (2000), Focusing of Laguerre-Gaussian beams by axicon, *Opt. Commun.*, 184, 105–112.

Johnson, S. L., P. Beaud, C. J. Milne, F. S. Krasniqi, E. S. Zijlstra, M. E. Garcia, M. Kaiser, D. Grolimund, R. Abela, and G. Ingold (2008), Nanoscale depth-resolved coherent femtosecond motion in laser-excited bismuth, *Phys. Rev. Lett.*, 100, 155501.

Joudkazis, S., H. Misawa E. Gamaly, B. Luther-Davies L. Hallo, P. Nicolai, and V. T. Tikhonchuk (2006a), Laser-induced micro-explosion inside of sapphire crystal: Evidence of multi-megabar pressure, *Phys. Rev. Lett.*, 96, 166101.

Joudkazis, S., H. Misawa, T. Hashimoto, E. Gamaly, and B. Luther-Davies (2006b), Laser-induced micro-explosion confined in a bulk of silica: Formation of nano-void, *Appl. Phys. Lett.*, 88(1), 201909.

Juodkazis, S., V. Mizekis, and H. Misawa (2009), Three-dimensional microfabrication of materials by femtosecond laser for photonics applications, *J. Appl. Phys.*, 106, 051101.

Juodkazis, S., A. Vailionis, E. G. Gamaly, L. Rapp, V. Mizeikis, and A. V. Rode (2016), Femtosecond laser-induced confined micro-explosion: Tool for creation high-pressure phases. *MRS Adv.*, Available on CJO 2015; doi:10.1557/adv.2015.41.

Kaganov, M. I., I. M. Lifshitz, and L. V. Tanatarov (1957), Relaxation between electrons and the crystalline lattice, *Sov. Phys. JETP*, 4(2), 173.

Kanter, H. (1970), Slow-electron mean free paths in aluminum, silver, and gold, *Phys. Rev. B*, 1, 522.

Kautek, W., and F. Krausz (1998), Femtosecond optical breakdown in dielectrics, *Phys. Rev. Lett*, 80, 4076–4079.

Keldysh, L. (1965), Ionization in the field of a strong electro-magnetic wave, *Sov. Phys. JETP*, 20, 1307.

Kim, Don H. Ehrenreich, and E. Runge (1994), Band structure of femtosecond-laser-pulse excited GaAs, *Solid State Commun.*, 89(2), 119–122.

Kittel, C. (1996), *Introduction to Solid State Physics*, Wiley & Sons.

Kruer, W. (1988), *The Physics of Laser Plasma Interactions*, Addison-Wesley.

Landau, L. D. (1937), Kinetic equation in the case of a Coulomb interaction, *Zh. Eksp.i Theor. Fiz.* (*Sov Phys. JETP*), 7, 203–209.

Landau, L. D., and E. M. Lifshitz (1980), *Statistical Physics*, Pergamon Press.

Landau, L. D., E. M. Lifshitz, and L. P. Pitaevskii (1984), *Electrodynamics of Continuous Media*, Pergamon Press.

Landolt-Börnstein (1983), *Numerical Data and Functional Relationships in Science and Technology, Group III, vol 17, Semiconductors,* ed. O. Madelung, M. Schulz, and H. Weiss (Springer-Verlag).

Lautenschlager, P., P. B. Allen, and M. Cardona (1985), Temperature dependence of band gaps in Si and Ge, *Phys. Rev. B*, 31, 2163–2171.

Laville, S., F. Vidal, T. W. Johnston, M. Haker, and B. Le Drogoff, O. Barthelemy, J. Argot, and M. Sabasabi (2004), Modelling the time evolution of laser-induced plasmas for various pulse durations and fluences, *Phys. Plasmas*, 5, 2182–2190.

Lennard-Jones, J. E., and A. F. Devonshire (1939), Critical and co-operative phenomena. III. A theory of melting and the structure of liquids, *Proc. Roy. Soc. A*, 169, 317.

Lenzner, M., J. Kruger, W. Kautek, and F. Krausz (1999), Incubation of laser ablation in fused silica with 5-fs pulses, *Appl. Phys. A*, 69, 465–466.

Lenzner, M., J. Kruger, S. Sartania, Z. Cheng, C. Spielmann, G. Mourou, W. Kautek, F. Krausz (1998), Femtosecond optical breakdown in dielectrics, *Phys. Rev. Lett.*, 80, 4076–4079.

Lifshitz, E. M., and L. P. Pitaevski (1981), *Physical Kinetics*, Pergamon Press.

Lindemann, F. (1910), The calculation of molecular vibration frequencies, *Physik. Zeits.*, 11, 609–612 (in German).

Liu, H., and P. Lalanne (2008), Microscopic theory of the extraordinary optical transmission, *Nature*, 452, 728–731.

Luo, S.-N., T. J. Arens, and P. D. Asimov (2003), Polymorphism, superheating and amorphization of silica upon shock wave loading and release, *J. Geophys. Res.*, 108, 2421.

Luther-Davies, B., E. G. Gamaly, Y. Wang, A. V. Rode, and V. T. Tikhonchuk (1992), Matter in ultra-strong laser fields, *Sov. J. Quantum Electron.*, 22, 289–325.

Luther-Davies, B., A. V. Rode, N. R. Madsen, and E. G. Gamaly (2005), Picosecond high-repetition rate pulsed laser ablation of dielectrics: The effect of energy accumulation between pulses, *Opt. Eng.*, 44(5), 051102.

MacDonald, W. M., M. N. Rosenbluth, and W. Chuck (1957), Relaxation of a system of particles with Coulomb interactions, *Phys. Rev.*, 107, 350–353.

Maine, P., D. Strickland, P. Bado, M. Pessot, and G. Mourou (1988), Generation of ultrahigh peak power pulses by chirped pulse amplification, *IEEE J. Quant. Electr.*, 24, 398–403.

Malinauskas, Mangirdas, Sima Reksˇtyte˙, Tomas Jonavicˇius, Vygantas Mizeikis, Eugene G. Gamaly and Saulius Juodkazis (2015), Nano-precision of 3D direct laser writing controlled by the coupling between tight focusing, polarization and heat conduction, *Laser Photonics Rev.*, March 22.

McMillan, P. F. (2002), New materials from high-pressure experiments. *Nat. Mater.*, 1, 19–25.

McMillan, P. F. (2005), Pressing on: The legacy of Percy W. Bridgman, *Nat. Mater.*, 4, 715–718.

Misawa, H., and Juodkazis, S., eds. (2006), *3D Laser Microfabrication: Principles and Applications*, Wiley-VCH Verlag Gmbh & Co.

Momma, C., S. Nolte, B. N. Chichkov, F. V. Alvensleben, A. Tunnermann (1997), Precise laser ablation with ultra-short pulses, *Appl. Surf. Sci.*, 109/110, 15–19.

Moore, D. S. (2018), Shock physics at the nanoscale [Invited], *J. Opt. Soc. Am. B*, 35; doi: 10.1364/JOSAB.35.0000B1.

More, R., K. Warren, D. Young, and G. Zimmerman (1988), A new quotidian equation of state (QEOS) for hot dense matter, *Phys. Fluids*, 31, 3059–3078.

Morse, P., and H. Feshbach (1953), *Methods of Theoretical Physics*, vols. 1, 2, McGraw Hill.

Mourou, G., Tajima, and Bulanov, S. (2006), Optics in the relativistic regime, *Rev. Mod. Phys.*, 78(2); doi: 10.1103/RevModPhys.78.309.

Mysyrowicz, A., R. Danylo, A. Houard, V. Tikhonchuk, X. Zhang, L. Yuan, and Y. Liu (2019), Lasing without population inversion in N_2^+, *APL Photon.*, 4, 110807; doi: 10.1063/1.5116898.

Nieminen, T. A., A. B. Stilgoe, N. R. Heckenberg, and H. Rubinsztein-Dunlop (2008), Angular momentum of a strongly focused Gaussian beam. *J. Opt. A: Pure Appl. Opt.*, 10, 115005.

Nikogosyan, D. (1988), *Properties of Optical and Laser-Related Materials, A Handbook*, John Wiley& Sons.

Nolte, S., C. Momma, H. Jacobs, A. Tünnermann, B. N. Chichkov, B. Wellegehausen, and H. Welling (1997), Ablation of metals by ultrashort laser pulses, *J. Opt. Soc. Am. B*, 14, 2716–2722.

Penn, D. (1980), Mean-free path of very low energy electrons: The effect of exchange and correlations, *Phys. Rev. B*, 22, 2677–2682.

Perelomov, A. M., V. S. Popov, and M. V. Terent'ev (1966), Ionization of atoms in an alternative electric field, *Sov. Phys. JETP*, 23(5), 924–934.

Perry, M. D., B. C. Stuart, P. S. Banks, M. D. Feit, V. Yanovsky, and A. M. Rubenchik (1999), Ultra-short-pulse laser machining of dielectric materials, *J. Appl. Phys.*, 85, 6803–6810.

Pickard, C. J., and Needs, R. J. (2010), Aluminium at terapascal pressures, *Nat. Mater.*, 9, 624–627.

Pickard, C. J., and R. J. Needs (2011), Ab initio random structure searching. *J. Phys. Condens. Matter*, 23, 053201.

Pines, D. (1964), *Elementary Excitations in Solids*, W. A. Benjamin, INC.

Porras, M. A., A. Parola, D. Faccio, A. Dubietis, and Paolo Di Trapani (2004), Nonlinear unbalanced bessel beams: Stationary conical waves supported by nonlinear losses, *Phys. Rev. Lett.*, 93, 153902.

Pronko, P. P., P. A. VanRompay, C. Horvath, F. Loesel, T. Juhasz, X. Liu, and G. Mourou (1998), Avalanche ionisation and dielectric breakdown in silicon with ultrafast laser pulses, *Phys. Rev. B*, 58, 2387–2390.

Prutton, M. (1994), *Introduction to Surface Physics*, Clarendon.

Quinn, J. J., and R. A. Ferrell (1958), Electron-self-energy approach to correlation in a degenerate electron gas, *Phys. Rev.*, 112, 812–827.

Rapp, L. , B. Haberl, C. J. Pickard, J. E. Bradby, E. G. Gamaly, J. S. Williams, and A. V. Rode (2015), Experimental evidence of new tetragonal polymorphs of silicon formed through ultrafast laser-induced confined microexplosion, *Nat. Commun.*, 6, 7555.

Raizer, Y. P. (1976), *Laser Spark and Propagation of Discharges*, Consultant Bureau.

Raizer, Y. P. (1977), *Laser-Induced Discharge Phenomena*, Consultant Bureau.

Rapp, L., R. Meyer, R. Giust, L. Furfaro, M. Jacquot, P. A. Lacourt, J. M. Dudley, and F. Courvoisier (2016), High aspect ratio micro-explosions in the bulk of sapphire generated by femtosecond Bessel beams, *Sci. Rep.*, 6, 34286; doi: 10.1038/srep34286.

Rapp, L., (2020), Private communication.

Reichert, M., Smirl, A. L., Salamo, G., Hagan, D. J., and E. W. van Stryland (2016), Observation of non-degenerate two-photon gain in GaAs, *Phys. Rev. Lett.*, 117, 73602.

Rethfeld, B., K. Sokolowski-Tinten, D. von der Linde, and S. I. Anisimov (2002), Ultra-fast thermal melting of laser-excited solid by homogeneous nucleation, *Phys. Rev. B*, 65, 092103-1-4.

Rode, A. V., D. Boschetto, T. Garl, and A. Rousse (2009), Transient dielectric function of fs-laser excited bismuth, in: *Ultrafast Phenomena XVI*, ed. P. Corkum, S. de Silvestri, K. A. Nelson, E. Riedle, and R. W. Schoenlein (Springer).

Rode, A. V., Luther-Davies, B., and Gamaly, E. G. (1999), Ultrafast ablation with high-pulse-rate lasers. Part II: Experiments on laser deposition of amorphous carbon films, *J. Appl. Phys.*, 85, 4222–4230.

Rosales-Guzmán, C., and A. Forbes (2017), *How to Shape Light with Spatial Light Modulators*, Bellingham (https://www.spiedigitallibrary.org/ebooks/).

Rousse, A., G. Rischel, S. Fourneaux, I. Uschmann, S. Sebban, G. Grillon, Ph. Balcou, E. Förster, J. P. Geindre, P. Audebert, J. C. Gauthier, and D. Hulin (2001), Non-thermal melting in semi-conductors measured at femtosecond resolution, *Nature*, 410, 65–68.

Rozmus, W., and V. T. Tikhonchuk (1990), Skin effect and interaction of short laser pulses with dense plasmas, *Phys. Rev. A*, 42, 7401–7412.

Rozmus W., and V. T. Tikhonchuk (1992), Heating of solid targets by subpicosecond laser pulses, *Phys. Rev. A*, 46, 7810–7814.

Salo, J., J. Fagerholm, A. T. Friberg, and M. M. Salomaa (2000), Unified description of nondiffracting X and Y waves, *PRE*, 62(3), 4261.

Salleo, A., S. T. Taylor, M. C. Martin, W. R. Panero, R. Jeanloz, T. Sands, and F. Y. Genin (2003), Laser-driven formation of a high-pressure phase in amorphous silica, *Nat. Mater.*, 2, 796–800.

Sanner, N., O. Uteza, B. Chimier, M. Sentis, P. Lassonde, F. Legare, and J. C. Kieffer (2010), Toward determinism in surface damaging of dielectrics using few-cycle laser pulses, *Appl. Phys. Lett.*, 96, 071111.

Schaffer, C. B., J. F. Garcia, and E. Mazur (2003), Bulk heating of transparent materials using a high-repetition rate femtosecond laser, *Appl. Phys. A*, 76, 351–354.

Sedov, L. I. (1959), *Similarity and Dimensional Methods in Mechanics*, Academic Press.

Silin, V. P. (1965), Nonlinear high-frequency plasma conductivity, *Zh. Eksp. Teor. Fiz.*, 47, 2254; (1964), *Sov. Phys. JETP*, 20, 1510.

Siwick, B. J., J. R. Dwyer, R. E. Jordan, and R. J. D. Miller (2003), An atomic level view of melting using femtosecond electron diffraction, *Science*, 302, 1382.

Smyth,J., S. Jacobsen, and R. Hanzen, (2000), *Comparative Crystal Chemistry of Dense Oxide Minerals*, vol. 41, Mineralogical Society of America, Washington, chap. 6, pp. 157–186.

Sokolowski-Tinten, K., C. Blome, J. Blums, A. Cavalleri, C. Dietrich, A. Tarasevitch, I. Uschmann, E. Foerster, M. Kammler, M. Horn-von-Hoegen, and D. von der Linde (2003), Femtosecond X-ray measurement of coherent lattice vibrations near the Lindemann stability limit, *Nature*, 422, 287.

Sokolowski-Tinten, K., and D. von der Linde (2000), Generation of dense electron–hole plasmas in silicon, *Phys. Rev. B*, 61, 2643–2650.

Sosman, R. B. (1965), *The Phases of Silica*, Rutgers University Press.

Sparks, M., D. L. Mills, R. Warren, T. Holstein, A. A. Maradudin, L. J. Sham, E. Loh, Jr., and D. F. King (1981), Theory of electron avalanche in solids, *Phys. Rev. B*, 24, 3519–3536.

Stuart, B. C., M. D. Feit, S. Herman, A. M. Rubenchik, B. W. Shore, and M. D. Perry (1996), Optical ablation by high-power short-pulse lasers, *J. Opt. Soc. Am. B*, 13, 459–468.

Stuart, B. C., M. D. Feit, A. M. Rubenchick, B. W. Shore, and M. D. Perry (1995), Laser-induced damage in dielectrics with nanosecond to picosecond pulses, *Phys. Rev. Lett.*, 74, 2248–2251.

Sun, H., Y. Xu, S. Juodkazis, K. Sun, M. Watanabe, S. Matsuo, H. Misawa, and J. Nishii (2001), Arbitrary-lattice photonic crystals created by multi-photon micro-fabrication, *Opt. Lett.*, 26(6), 325.

Swift, D. C., et al. (2012), Gigabar material properties experiments on NIF and OMEGA, *AIP Conf. Proc.*, 1426, 477–480.

Tallon, J. L. (1984), Communal entropy in melting and the glass, fast-ion, and superfluid transitions, *Phys. Rev. B*, 29, 4153–4155.

Tallon, J. L. (1989), A hierarchy of catastrophes as a succession of stability limits for the crystalline state, *Nature*, 342, 658–660.

Tamm, I. E. (1976), *Basics of the Theory of Electricity*, 9th edition, Nauka (in Russian).

Tateno, S., K. Hirose, Y. Ohishi, and Tatsumi, Y. (2010), The structure of iron in earth's inner core, *Science* 330, 359–361.

Temnov, V. V., K. Sokolowski-Tinten, P. Zhou, A. El-Khamhawy, and D. von der Linde, (2006), Multiphoton ionization in dielectrics: Comparison of circular and linear polarization, *Phys. Rev. Lett.*, 97, 237403.

Tien, A.-C., S. Backus, H. Kapteyn, M. Murname, and G. Mourou (1999), Short-pulse laser damage in transparent materials as a function of pulse duration, *Phys. Rev. Lett.*, 82, 3883–3886.

Tikhonchuk, V. T., (2020), Private communication.

Trunin, R. F. (1994), Shock compressibility of condensed materials in strong shock waves generated by underground nuclear explosions, *Phys. Usp.*, 37, 1123–1146.

Uhlenbeck,G. E., and L. S. Ornstein, (1930), On the theory of the Brownian motion, *Phys. Rev.*, 36, 823–841.

Uteza, O. P., E. G. Gamaly, A. V. Rode, M. Samoc, and B. Luther-Davies (2004), Gallium transformation under femtosecond laser excitation: Phase coexistence and incomplete melting, *Phys. Rev. B*, 70, 054108.

Uteza, O., N. Sanner, B. Chimier, A. Brocas, N. Varkentina, M. Sentis, P. Lassonde, F. Legare, and J. C. Kieffer (2011), Control of material removal of fused silica with single pulses of few optical cycles to sub-picosecond duration *Appl. Phys. A*, 105, 131–141.

Vailionis, A., E. G. Gamaly, V. Mizeikis, Wenge Yang, A. V. Rode, and S. Juodkazis (2011), Discovery of super-dense Al: Synthesis by ultrafast microexplosion, *Nat. Commun.*, 2, 445; doi: 10.1038/ncomms1449.

Vidal, F., T. W. Johnston, S. Laville, O. Barthelemy, M. Chaker, B. Le Drogoff, J. Margot, and M. Sabsabi (2001), Critical point phase separation in laser ablation of conductors, *Phys. Rev. Lett.*, 86, 2573–2576.

Vijayakumar, A., J. Maksimovic, T. Katkus, S. H. Ng, O. Ulcinas, M. Mikutis, J. Baltrukonis, A. Urbas, G. Slekys, H. Ogura, D. Sagae, T. Pikuz, T. Somekawa, N. Ozaki, A. Vailionis, G. Seniutinas, V. Mizeikis, K. Glazebrook, J. P. Brodie, P. R. Stoddart, L. Rapp, A. V. Rode, E. G. Gamaly, S. Juodkazis (2020), All femtosecond optical pump and X-ray probe: Holey-Axicon for free electron laser, https://www.preprints.org/manuscript/202005.0416/v1.

Vinogradov, A. V. (1975), Absorption of high-intensity light by free carriers in dielectrics, *Sov. Phys. JETP*, 41(3), 540–543.

Wang, G., Y. Yu, L. Jiang, X. Li, Q. Xie, and Y. Lu (2017), Cylindrical shockwave-induced compression mechanism in femtosecond laser Bessel pulse micro-drilling of PMMA, *Appl. Phys. Lett.*, 110, 161907.

Weast, R. C., and M. J. Astle, eds. (1981), *CRC Handbook of Chemistry and Physics,* 60th edition, CRC Press.

Winkler, T., L. Haahr-Lillevang, C. Sarpe, B. Zielinski, N. Götte, A. Senftleben, P. Balling, and T. Baumert (2017), Laser amplification in excited dielectrics, *Nat. Phys.*, 14(1), 74–79.

Wood, W. M., G. Focht, and M. C. Downer (1988), Tight focusing and blue shifting of millijoule femtosecond pulses from a conical axicon amplifier, *Opt. Lett.*, 13, 984.

Yablonovitch, E., and N. Bloembergen (1972), Avalanche ionization and the limiting diameter of filaments induced by light pulses in transparent media, *Phys. Rev. Lett.*, 29, 907–910.

Zangwill, A. (1988), *Physics at Surfaces,* Cambridge University Press.

Zegriotti, I., S. Mailis, N. A. Vainos, P. Papakostantinou, C. Kalpouzos, C. P. Grogoropoulos, and C. Fotakis (1998), Micro-depositionof metal and oxide structures using ultr-short laser pulses, *Appl. Phys. A*, 66, 579.

Zeiger, H. J., J. Vidal, T. K. Cheng, E. P. Ippen, G. Dresselhaus, and M. S. Dresselhaus (1992), Theory for displacive excitation of coherent phonons, *Phys. Rev. B*, 45, 768.

Zel'dovich, Ya. B., and Y. P. Raizer (2002), *Physics of Shock Waves and High-Temperature Hydrodynamic Phenomena*, Dover.

Index